# Ordered Banach spaces



*Collection Travaux en cours*, 68

COLECTION TRAVAUX EN COURS

| | | | |
|---|---|---|---|
| 1 | **J. Vaillant I.** Équations aux dérivées partielles hyperboliques et holomorphes | 2 7056 5962 5 |
| 2 | **J. Mossino** Inégalités isopérimétriques et applications en physique | 2 7056 5963 3 |
| 3 | **A. Benabdallah** Opérateurs différentiels invariants | 2 7056 5964 1 |
| 4 | **B. Beauzamy, J.T. Lapresté** Modèles étalés des espaces de Banach | 2 7056 5965 x |
| 5 | **P. Dazord, N. Desolneux-Moulis** Séminaires sud-rhodaniens de géométrie ( 6 vol.) | |
| | I. Géométrie symplectique et de contact | 2 7056 5971 4 |
| 6 | **II.** Feuilletages et quantification géométrique | 2 7056 5972 2 |
| 7 | **III.** Autour du théorème de Poincaré-Birkhoff | 2 7056 5973 0 |
| 8 | **J. Bernstein:** Repérsentations des groupes réductifs sur un corps local | 2 7056 5978 1 |
| 9 | **C. Albert, P. Molino** Pseudogroupes de Lie transitifs | |
| | **I.** Structures principales | 2 7056 5989 7 |
| 10 | **M. Zamansky** Analyse harmonique et approximation | 2 7056 6002 x |
| 11 | **J. Vaillant** Équations aux dérivées partielles (3 vol.) | |
| | **II.** Propagation des singularités et opérateurs différentiels | 2 7056 6003 8 |
| 12 | **J.P. Dufour** Séminaires sud-rhodaniens de géométrie (2 vol.) | |
| | **IV.** Géométrie symplectique et mécanique | 2 7056 6004 6 |
| 13 | **V.** Singularités, feuilletages et mécanique hamiltonienne | 2 7056 6005 4 |
| 14 | **V. Avanissian** Cellule d'harmonicité et prolongement analytique complexe | 2 7056 6012 7 |
| 15 | **M. Zamansky** Approximation des fonctions | 2 7056 6013 5 |
| 16 | **A. Pajor** Sous-espaces ln1 des espaces de Banach | 2 7056 6021 6 |
| 17 | **J.L. Mauclaire** Intégration et théorie des nombres | 2 7056 6035 6 |
| 18 | **T. Ekedahl** Diagonal complexe and F-Gauge structures | 2 7056 6045 3 |
| 19 | **C. Albert, P. Molino** Pseudogroupes de Lie transitifs (2 vol.) | |
| | **II.** Théorème d'intégrabilité | 2 7056 6055 0 |
| 20 | **M. Audin** Cobordismes d'immersions lagrangiennes et legendriennes | 2 7056 6056 9 |
| 21 | **Y. Choquet-Bruhat, A. Lichnerowicz** Géométrie et physique | 2 7056 6057 7 |
| 22 | **J.M. Aroca, T. Sanchez-Giralda.** Géométrie algébrique et applications (3 vol.) | |
| | **I.** Géométrie et calcul algébrique (épuisé) | 2 7056 6027 5 |
| 23 | **II**. Singularités et géométrie complexe (épuisé) | 2 7056 6028 3 |
| 24 | **III.** Géométrie réelle (épuisé) | 2 7056 6029 1 |
| 25 | **VI** Aspects dynamiques et topologiques des groupes infinis | 2 7056 6062 3 |
| 26 | **VII** Feuilletages riemanniens, quant. géométriques et mécaniques | 2 7056 6076 3 |
| 27 | **VIII**. Actions hamiltoniennes de groupe. 3e théorième de Lie | 2 7056 6085 2 |
| 28 | **J.M. Ghidaglia,** Équations aux dérivées partielles non linéaires dissipatives | |
| | et systèmes dynamiques (épuisé) | 2 7056 6084 4 |
| 29 | **J. Vaillant** Équations aux dérivées partielles | |
| | **III.** Calcul d'opérateurs et fronts d'ondes | 2 7056 6078 x |
| 30 | **D. Kastler** Cyclic cohomology within the differential envelope | 2 7056 6044 5 |
| 32 | **D. Bernard, Y. Choquet-Bruhat** I. Physique quantique et géométrie (épuisé) | 2 7056 6087 9 |
| 33 | **D. Bernard, Y. Choquet-Bruhat II.** Géométrie différentielle (épuisé) | 2 7056 6088 7 |
| 34 | **Lê Dung Trang** Introduction à la théorie alg. des systèmes différentiels | 2 7056 6048 8 |
| 35 | **Z. Mebkhout** Systèmes différentiels. 6 opérations de Grothendieck (épuisé) | 2 7056 6049 6 |
| 36 | **Lê Dung Trang I** Singularités et monodromie | 2 7056 6091 7 |
| 37 | **Lê Dung Trang II** Méthodes algébriques et géométriques | 2 7056 6092 5 |
| 38 | **B. Angéniol,** Calcul différentiel et classes caractéristiques | 2 7056 6108 5 |
| 39 | **M.H. Schwartz** Champs radiaux sur une stratification analytique | 2 7056 6160 3 |

40 **A. Mortajine** Classification des espaces préhomogènes 2 7056 6165 4
41 **R. Haydon,** Randomly normed spaces 2 7056 6180 8
42 **Shih Wei Hui** Sol. analytiques des équations aux dérivées partielles 2 7056 6192 1
43 **G. Roos, J.P. Vigué** Systèmes triples de Jordan. Domaines symétriques 2 7056 6205 7
44 **H. Rubenthaler** Algèbres de Lie et espaces préhomogènes 2 7056 6206 5
45 **P. Maisonobe, C. Sabbah** D-modules cohérents et holonomes 2 7056 6212 x
46 **P. Maisonobe, C. Sabbah** Images directes et constructibilité 2 7056 6213 8
47 **L. Boutet de Monvel** Analyse algébrique des perturbations singulières (2 vol.)
**I.** Méthodes résurgentes 2 7056 6229 4
48 **II.** Méthodes différentielles 2 7056 6230 8
49 **E. Sanchez-Palencia** Spectral analysis of complex structures 2 7056 6262 6
50 **M. Goze** Anneaux et modules 2 7056 6279 0
51 **M. Goze** Lois d'algèbre et variétés algébriques 2 7056 6280 4
52 **J.M. Gambaudo, R. Bamon, S. Martinez** Dynamical systems 2 7056 6288 x
53 **J.M. Gambaudo, R. Bamon, S. Martinez** Disordered systems 2 7056 6289 8
54 **D. Trotman, L.C. Wilson** Stratifications, singularities and differential equations (2 vol.)
**I.** Singularities of maps 2 7056 6318 5
55 **II.** Stratification and topology2 7056 6319 3
56 **M. Boileau, M. Domergue, Y. Mathieu, K. Millet** Progress in knot theory 2 7056 6334 7
57 **P. Mardesic** Chebyshev syst. and the versal unfolding of the cusps of order n 2 7056 6359 2
58 **M. Weber** Entropie métrique et convergence presque partout 2 7056 6381 9
59 **R. Becker** Cônes convexes en analyse 2 7056 6384 3
60 **G. Maugin** Geometry, continua and microstructure 2 7056 6399 1
61 **G. Schwarz, M. Brion** Théorie des invariants et géo. des variétés quotients 2 7056 6405 x
62 **P. Orro, F. Pelletier** Singularités et géométrie sous-riemanienne 2 7056 6422 x
63 **V. Lakshmibai, N. Gonciulea** Flag varieties 2 7056 6389 4
64 **T. Sari** Contrôle non linéaire et applications 2 7056 6511 0
65 **E.Pécou, B. Prum** Mathematical and computational methods in biology 2 7056 6578 1
66 **A. Mass, S. Martinez,J. San Martin** Information and Randomness 2 7056 6578 1
67 **R. Becker** Convex Cones in analysis 2 7056 6671 0
68 **R. Becker** Ordered Banach spaces 2 7056 6721
69 **Z. Denkowska** Ensembles sous-analytiques à la polonaise 2 7056 6688 0

Richard Becker

# Ordered Banach spaces

*préface de Gilles Godefroy*

COLLECTION TRAVAUX EN COURS

Hermann Éditeurs des sciences et des arts

Isbn 978 27056 6721 4

# Contents

# Preface

Each time we meet and use an algebraic structure (group, field, graph, convex body, ...), we are bound to consider those subsets which enjoy the relevant property. And dealing with a vector space makes little sense unless its vector subspaces are part of the game. Weaker stability properties turn out to be important as well: when vector spaces are concerned, this is the case of convex cones. In other words, of those subsets which are stable under all linear combinations with positive coefficients. Their study and use is a significant part of functional analysis. But to remain on the algebraic side, one can observe that such a cone, when proper, defines an order structure which is compatible with addition and multiplication of vectors by a positive scalar.

Richard Becker's work, as displayed in the present book, originates from three different sources. Conical measures rank first. These objects were defined by Gustave Choquet for developing integral representation theory in cones with no compact base. Richard Becker, who originally was G. Choquet's PhD student (as I myself was), is an undisputed expert of this theory which can be described as "integral representation with as little compactness as possible" and naturally requests a careful analysis of convex cones. A second source of inspiration is the well-established theory of Banach lattices, or in other words the theory of complete function spaces where the norm of a function depends only on its absolute value. Such Banach spaces enjoy a quite simpler structure than general Banach spaces, although studying their algebra of operators lead to hard work, to fundamental results such as Bernard Maurey's factorization theorems, and also to open questions. The corresponding cones are named lattice cones (whose bases, if any, are simplices) but Richard Becker makes it clear that he considers in this work the more general normal cones, and even arbitrary ones. Last but not least, the third root of this work is Jean-Louis Krivine's theorem of finite representability of finite dimensional $l_p$ spaces into Banach lattices, which played a crucial role in local theory of Banach spaces and turned out to be optimal in its generality.

Collecting these bricks, investigating their links and fixing them with the classical tools of functional analysis such as Hahn-Banach theorem (under the disguise of Pietsch's factorization procedure) provides an original and instructive contribution to classical functional analysis by Richard Becker, who remains faithful to Gustave Choquet's word of advice: simplicity.

Gilles GODEFROY

# Introduction

**A)** This book is devoted to the theory of general *ordered Banach spaces.*

Extensive studies have been made of Banach spaces since 1970; among the books concerning this theory let us mention P. Wojtaszczyk [68] (in 1991), J. Diestel, H. Jarchow and A. Tonge [20] (in 1995), the Handbook (in 2001), D. Li and H. Queffelec [41] (in 2004) and F. Albiac and N.J. Kalton [2] (in 2006). However, these books do not deal with general ordered Banach spaces at all, except for the Handbook [1] (see the Article by Y.A. Abramovich and C.D. Aliprantis p. 87-88). Some others are devoted to a specific case of ordered Banach spaces, the one of *Banach lattices*: see for instance H.H. Schaefer [63] (in 1974), C.D. Aliprantis and O. Burkinshaw [3] (in 1985), P. Meyer-Nieberg [48] (in 1991) and the aforementioned reference [1].

Notice that, however, if $E$ and $F$ are two Banach lattices, the cone of all *positive operators* from $E$ to $F$ is not necessarily a *lattice cone*; this is one important reason (among others) to investigate general ordered Banach spaces. Another interesting example is the case of Sobolev's spaces.

The aim of the present book is to fill this gap and to provide an up to date study of the structure of *convex cones* contained in a Banach space and of some related *operators.*

Between 1962 and 1992 this topic was more or less dormant; several books describe what was known before 1962, for example those of I. Namioka [49] (in 1957), G. Jameson [30] (in 1970) and H.H. Schaefer [62] in 1971.

I think that this book is accessible to any student with a background in Functional Analysis consisting of elementary Banach spaces theory and of measure theory. As J. Von Neumann used to say, "I have had no time to make things difficult". In fact, I have spent a lot of time on clarifying the main points and keeping the book self-contained.

On the other hand, any reader involved in the theory of Banach Spaces and *Summing Operators* will realize that *Grothendieck's Inequality*, which plays a fundamental role in Diestel's book [20], is not drawn on here nor is there much concerning B. Maurey and G. Pisier's *Type* and *Cotype* theory (see Appendix 1).

However, there is one drawback in this presentation: a priori, if $B$ is a Banach space there is no reason to investigate the *factorization* of operators from $B'$ to a space $L^1$ through $L^p$ (as in Chapter II); the only reason for this, but it is a paramount one, is that it works !

Finally, I hope that this book will encourage the reader to solve the problems we mention and so that this book will appear very incomplete before long.

**B)** To begin with, let us explain what an *ordered Banach space* is:

If $A$ is a set, an *order* on $A$ is a *binary* relation $x \geq y$ which is *reflexive* ($x \geq x$ for every $x \in A$), *transitive* ($x \geq y$ and $y \geq z$ imply $x \geq z$) and *anti-symmetric* ($x \geq y$ and $y \geq x$ imply $x = y$); when the last condition is omitted, $\geq$ is said to be a *preorder.*

If $A$ is equipped with an order, $A$ is said to be a *lattice* if for all $x, y \in A$ the elements $x \vee y = \sup\{x, y\}$ and $x \wedge y = \inf\{x, y\}$ do exist in $A$.

If $V$ is a real vector space, $V$ is said to be a *preordered vector space* if $V$ is equipped with a preorder denoted by $\geq$ such that:

(Or 1): If $x, y \in V$ satisfy $x \geq y$, for any $z \in V$ one has $x + z \geq y + z$.
(Or 2): If $x \in V$ satisfies $x \geq 0$, then for any scalar $\lambda \geq 0$ one has $\lambda x \geq 0$.

By axiom (Or 1) the set $X = \{x : x \in V, x \geq 0\}$ is stable under finite sums and is even a *convex cone* by (Or 2). $X$ is called the *positive cone* of $V$ and its elements are called *positive.*

By axiom (Or 1) one has $x \geq y$ iff $x - y \in X$.

It is clear that the relation $\geq$ is an *order* iff $X$ is a *proper* cone, namely if $X \cap -X = \{0\}$.

Conversely, it is easy to see that, if $X$ is any convex cone contained in a real vector space $V$, the relation $x \geq y$ iff $x - y \in X$ defines a preorder on $V$ satisfying the above two axioms. $X$ is said to be the *positive cone* of $V$.

Here is a particular case: if $V$ is an ordered vector space, we say that $V$ is a *vector lattice* if $V$ is a lattice for this order; we set $|x| = x \vee -x$. The positive cone of $V$ is then said to be a *lattice cone.* It is easy to see that in $\mathbb{R}^2$ every closed proper convex cone with non-empty interior is a lattice cone.

Of course, everybody knows what a Banach space is; this book shall only deal with *real* Banach spaces. A *preordered* Banach space is a Banach space which is a preordered vector space.

Consequently, in a way, the study of preordered Banach spaces amounts to the study of convex cones contained in Banach spaces. When the cone is the whole Banach space we are in the framework of the classical Banach spaces theory.

Here are some examples of ordered Banach spaces:

A particular case is the one of *Banach lattices*; namely, an ordered Banach space $B$ is a *Banach lattice* if $B$ is a vector lattice such that for all $x, y \in B$ with $|x| \leq |y|$ one has $\|x\| \leq \|y\|$.

Every $L^p$ space for $p \geq 1$ is a Banach lattice for the classical norm; the same thing is true for $\ell^p$ spaces. If $\Omega$ is a Hausdorff topological space the space $\mathcal{C}_b(\Omega)$ of all real continuous bounded functions on $\Omega$ is a Banach lattice for the sup-norm.

Banach lattices have been studied in considerable depth; a classical reference is the book by H.H. Schaefer [63] (in 1974); a more recent reference (1991) which is probably the most complete one is the book by P. Meyer-Nieberg [48].

But many ordered Banach spaces are not Banach lattices, even if they are vector lattices; here is an example:

Let $B = H_0^1([0,1])$ be the Sobolev space consisting of all *absolutely continuous* real functions $f$ defined on $[0,1]$ such that $f(0) = f(1) = 0$ and that $\int_0^1 (f'(u))^2 du < \infty$ equipped with the norm $\|f\| = (\int_0^1 (f'(u))^2 du)^{1/2}$. It is well known that $B$ is a Banach space (and even a Hilbert space) and it is clear that $B$ is a vector lattice for the classical order of functions because if $f, g \in B$ one has $f \vee g \in B$; but $B$ is not a Banach lattice: indeed, let $f \in B$ be any non-zero element; for every integer $n$ it is easy to find $f_n \in B$ such that $|f_n| \leq |f|$ and that $\|f_n\| > n$.

Here is another example: if $E$ and $F$ are two Banach lattices the convex cone of all continuous operators $T : E \to F$, which are *positive* (i.e. $x \geq 0$ implies $T(x) \geq 0$) is generally not a lattice cone.

In other words it is quite natural to study ordered Banach spaces which are not Banach lattices.

**C)** Up to 1992, only three main results concerning ordered Banach spaces, due to M.G. Krein [33] (in 1940), V.L. Klee [31] (in 1955) and T. Ando [4] (in 1962), were known:

In an article published in French in 1940 M.G. Krein [33] introduced the fundamental notion of *normal cones*.

1. Definition.

Let $X$ be a convex cone contained in a Banach space $B$. We say that $X$ is a *normal cone* if there exists a constant $C$ such that for all $x, y \in X$ one has $\|x\| \leq C\|x + y\|$.

We say that $X$ is *generating* if $B = X - X$.

Of course a normal cone $X$ is *proper*, namely $X \cap -X = \{0\}$.

The positive cone of a Banach lattice is normal and generating; the convex cone of all positive operators from a Banach lattice to another one (for the operator norm) is normal; the positive cone of the Sobolev space $H_0^1([0,1])$ mentioned above is generating but not normal.

If $B$ is a Banach space we denote by $B'$ its dual. If $X \subset B$ is a convex cone (not necessarily proper) let $X^0$ be the polar cone of $X$ in $B'$; namely

$$X^0 = \{\ell : \ell \in B', \ell(x) \geq 0 \text{ for all } x \in X\}$$

Here is the theorem proved in 1940 by M.G. Krein [33]:

2. Theorem.
If $X \subset B$ is a normal cone, then $X^0$ is a generating cone in $B'$.

In 1955 V.L. Klee [31] proved the following:

3. Theorem.
If $X \subset B$ is a closed convex generating cone (not necessarily proper), then every $x \in B$ admits a decomposition $x = x_1 - x_2$ where $x_1, x_2 \in X$ satisfy $\|x_1\| + \|x_2\| \leq C\|x\|$, where $C$ is some constant (independant of $x$).
Moreover, in this case, every linear form $f$ defined on $B$ which is positive on $X$ is continuous.

In 1962 T. Ando [4] proved the following:

4. Theorem.
Let $X \subset B$ be a closed convex cone; if $X^0$ is a normal cone in $B'$, then $X$ is generating in $B$.

In 1992 ([7] – [13]) I began to study ordered Banach spaces and more generally convex cones contained in a Banach space. Three *Deus ex machina* enabled me to carry out this study:
a) The theory of *factorization of operators* (see Chapter II), due to B. Maurey [44], from a Banach space to a space $L^1(m)$ through the space $L^p(m)$, where $p > 1$ and, in particular, a lemma due to H.P. Rosenthal (see II.8), appropriately extended and used by B. Maurey, which enabled him to carry out the "finer analysis" contained in his book, as he wrote himself ([44], p. 6).
As far as I know, this lemma has never been used since.
b) The theory of *conical measures* (see Chapter III) created by G. Choquet in the framework of the theory of *integral representation*, which will be useful here, for a quite different purpose.
c) An important theorem due to J.L. Krivine [35] (see V.11) concerning finite dimensional subspaces of a Banach lattice. A miraculous trick (see V.12) enables us to use this theorem in the framework of normal cones.

**D)** Let us now describe the Chapters of our book:

Chapter I: We first recall with sketched proofs various forms of Hahn-Banach's Theorem. We then prove the results of M.G. Krein, V.L. Klee and T. Ando mentioned above.

Chapter II: Let $E$ be a Banach space and $u$ be a continuous linear map from $E$ to a space $L^1(\Omega, \mathcal{A}, m)$, (where $\Omega$ is a set, $\mathcal{A}$ a $\sigma$-algebra of subsets of $\Omega$ and $m$ a positive $\sigma$-additive measure on $\mathcal{A}$); for any $1 < p < \infty$ we give, after B. Maurey [44], a necessary and sufficient condition ensuring that $u$ *factors* through $L^p(\Omega, \mathcal{A}, m)$ in the following sense:

There exist $g \in L^{p'}(\Omega, \mathcal{A}, m)$ $(1/p + 1/p' = 1)$ and some constant $C$ such that, for all $x \in E$, one has:

$$\|u(x)/g\|_p \leq C.\|x\|$$

We prove the Rosenthal Lemma mentioned above.

Chapter III: To study convex cones, it is necessary to deal with positive *linear forms* on some vector spaces consisting of functions defined on those cones, which are *positively homogeneous* of degree 1. This is why we define and study *conical measures*:

After G. Choquet, if $E$ is a Hausdorff locally convex space (H.l.c.s.) with continuous dual $E'$ a positive *conical measure* on $E$ is a positive linear form on the vector lattice of functions on $E$ generated by $E'$.

Chapter IV: This Chapter contains our first main results concerning the structure of convex cones (proper or not) contained in a Banach space: to every convex cone $X$ (not necessarily proper) contained in a Banach space we associate an *index* $i(X)$ in Definition 7 below. Before giving a summary of the results contained in this Chapter we need some notations and definitions.

5. Definition.

If $(x_n)$ is a sequence in a Banach space $B$ and if $p \geq 1$ we set:

$$M_p((x_n)) = \sup\{(\sum |x'(x_n)|^p)^{1/p} : x' \in B', \|x'\| \leq 1\}$$

If $M_p((x_n)) < \infty$ the sequence $(x_n)$ is said to be *scalarly* $\ell^p$.

6. Definition.

Let $X \subset B$ be a convex cone (not necessarily proper) and $p \geq 1$; a linear operator $T$ (not necessarily continuous), from $B$ to another Banach space $F$, is said to be *p-summing on X* if for some $C > 0$ one has, for every finite sequence $x_1, , , x_n$ in $X$:

$$(\sum_1^n \|T(x_i)\|^p)^{1/p} \leq CM_p((x_i))$$

We denote by $\pi_p(X, T)$ the minimal constant $C$ and by $\Pi_p(X; B, F)$ the set of these operators.

Here is the definition of the index just announced:

7. Definition

If $X \subset B$ is a convex cone (not necessarily proper) we set:

$$i(X) = \inf\{q : q \geq 1, \ \exists C, \ (\sum \|x_i\|^q)^{1/q} \leq CM_1((x_i))\}$$

for every finite sequence $x_1, , , x_n$ in $X$.

If $1 \leq r \leq \infty$ then $r'$ is defined by $1/r + 1/r' = 1$.

This index $i(X)$ is used all along our book.

Roughly speaking we prove that:

If $T$ is any operator from $B$ to another Banach space, which is $p$-summing on $X$, where $1 < p < i(X)'$, then $T$ is also 1-summing on $X$, and this is not always the case if $p > i(X)'$.

For every $1 < p < i(X)'$, every operator from $B'$ to a space $L^1$, which is positive on the polar cone $X^0 \subset B'$ of $X$, factors through $L^p$ (in the sense of Chapter II), and this is not always the case if $p > i(X)'$.

It is in this Chapter that we use the lemma of Rosenthal mentioned above and proved in II.8.

Chapter V: In Banach spaces theory a classical trend is to investigate finite dimensional $\ell^p$ spaces contained in a given Banach space. We deal with the case of normal cones; one of the main results states, roughly speaking, that, if $X \subset B$ is a normal cone such that $i(X) > 1$, then $i(X)$ is the *maximun* of all $r \geq 1$ such that $B$ contains uniform copies of spaces $\ell_n^r$ whose basis vectors are contained in $X$. We deduce that if $i(X) > 1$ then $p_B \leq i(X) \leq q_B$, where $p_B$ and $q_B$ are the classical *type index* and *cotype index* of $B$ (see Appendix 1). In particular, in a Hilbert space, any normal cone $X$ satisfies $i(X) = 1$ or $i(X) = 2$.

We study the particular case when $X$ is also generating.

It is in this Chapter that we use the theorem of J.L. Krivine (see V.11) mentioned above.

Chapter VI: Let $X \subset B$ be a convex cone (not necessarily proper); for any compact Hausdorff topological space $K$, if $u : \mathcal{C}(K) \to B$ is positive, in the sense that $u(f) \in X$ for every $f \geq 0$, we prove that $u$ is $r$-summing for every $r > i(X)$. When $X = B$ this theorem is due to B. Maurey.

We give applications to conical measures and to embeddings of ordered Banach spaces into a space $L^1(P)$, where $P$ is a probability.

Chapter VII: In the preceding Chapters we have encountered $p$-summing operators ($p \geq 1$) on a convex cone (not necessarily proper) contained in a Banach space; in this Chapter we study in depth this class of operators: we prove a *Pietsch* Inequality, we prove a *Composition Theorem*, we investigate the *transposed maps*. When the cone involved is the whole Banach space these results are classical.

The results of this Chapter will be useful in the next Chapter.

Chapter VIII: We describe a modification of the Lapreste *tensor norms* ([37]) between two Banach spaces to deal with the case when the Banach spaces involved are equipped with closed convex cones. The results obtained in the preceding Chapters enable us to compare the topologies so defined. Finally, we characterize some classes of operators between Banach spaces:

If $U$ and $V$ are two Banach spaces, and $\Gamma \subset V$ is a closed normal cone, for each $1 < s < \infty$ we determine the structure of all linear operators $\psi : U \to V$ such that, for every $L^{s'}$ space $B$ and every positive map $T : V \to B$, then $T \circ \psi$ is an *integral* map: namely we prove that this is equivalent to require that there exist a reflexive Banach space $W$ and $\alpha : U \to W$ and $\beta : W \to V$ such that $\psi = \beta \circ \alpha$ where $\alpha$ is $s$-summing and $\beta'$ is $s'$-summing on the polar cone $\Gamma^0 \subset V'$.

Five Appendices complete the book to make it self contained as much as possible, to raise problems and to relate how our results were obtained.

I would like to thank very much Mathieu Meyer for his helpful comments.

Richard BECKER

# Chapter I

# The state of the Art up to 1962

We state various versions of Hahn-Banach's Theorem (with sketched proofs) which will be useful in this book; then we prove the results of M.G. Krein, V.L. Klee and T. Ando mentioned in the Introduction (2-3-4).

In 1932, in his book, S. Banach [6] proved the following Theorem:

## I.1 Theorem

Let $V$ be a real vector space and $p : V \to \mathbb{R}$ be a sublinear form, namely $p$ satisfies $p(x+y) \leq p(x)+p(y)$ for all $x, y \in V$ and $p(\lambda x) = \lambda p(x)$ for all $x \in V$ and $\lambda \geq 0$. If $f$ is a linear form defined on a vector subspace $W \subset V$ such that $f(x) \leq p(x)$ for every $x \in W$, then $f$ has a linear extension $\widehat{f}$ to $V$ such that $\widehat{f}(x) \leq p(x)$ for every $x \in V$.

**Proof** (sketch)

Let $x_0 \in V \setminus W$; there exist a linear extension $f_0$ of $f$ to the vector space $W_0$ generated by $W$ and $x_0$ satisfying $f_0(x) \leq p(x)$ for every $s \in W_0$ iff:

1) $f_0(y + x_0) \leq p(y + x_0)$ for every $y \in W$ or equivalently $f_0(x_0) \leq p(y + x_0) - f(y)$ for every $y \in W$.

2) $f_0(z - x_0) \leq p(z - x_0)$ for every $z \in W$ or equivalently $f_0(x_0) \geq -p(z - x_0) + f(z)$ for every $z \in W$.

Consequently, the existence of $f_0$ is equivalent to the fact that there exists $a \in \mathbb{R}$ satisfying for all $y, z \in W$:

$$-p(z - x_0) + f(z) \leq a \leq p(y + x_0) - f(y)$$

This is equivalent to have for all $y, z \in W$: $-p(z-x_0)+f(z) \leq p(y+x_0) - f(y)$; this is equivalent to $f(y + z) \leq p(y + x_0) + p(z - x_0)$ for all $y, z \in W$.

But, by hypothesis, $f(y+z) \leq p(y+z)$ and $p(y+z) \leq p(y+x_0)+p(z-x_0)$; the result follows.

To conclude the proof it suffices to use Zorn's Lemma:

Let $\mathcal{Z}$ be the set of all pairs $(E, f_E)$ where $E$ is a vector subspace of $V$ with $W \subset E$ and $f_E$ is a linear form on $E$ satisfying $f_E = f$ on $W$ and $f_E(x) \leq p(x)$ for every $x \in E$.

$\mathcal{Z}$ is ordered by:

$$(E, f_E) \prec (F, f_F) \text{ if } E \subset F \text{ and } f_E = f_F \text{ on } E$$

By Zorn's Lemma $\mathcal{Z}$ has a *maximal* element (for the order $\prec$) denoted $(Z, f_Z)$; by the beginning of our proof (with $Z$ and $f_Z$) it follows that $Z = V$ and that $\widehat{f} = f_Z$ is convenient.

Five years before, in 1927, H. Hahn [28] (and independently in 1929 S. Banach [5]) had proved the following theorem (with the same proof as in Theorem I.1) which is an obvious consequence of the preceding:

## I.2 Theorem

Let $B$ be a Banach space; if $f$ is a linear form defined on a vector subspace $W \subset B$ such that $|f(x)| \leq M\|x\|$ for $x \in W$ then $f$ has a linear extension $\widehat{f}$ to $B$ such that $|\widehat{f}(x)| \leq M\|x\|$ for every $x \in B$.

In 1933, S. Mazur [47] proved the following result; we reproduce his proof. Note that a t.v.s. (topological vector space) is not necessarily locally convex.

## I.3 Theorem

Let $V$ be a t.v.s. not necessarily Hausdorff; if $A \subset V$ is a non-empty open convex set and if $M \subset V$ is an affine subspace such that $M \cap A = \emptyset$, there exists a closed hyperplane $H \subset V$ such that $M \subset H$ and $H \cap A = \emptyset$.

**Proof** (sketch)

By a translation we can assume that $0 \in A$; let $p : V \to \mathbb{R}$ be the gauge of $A$ defined by $p(x) = \inf\{r : r > 0, x \in rA\}$. Let $M_0$ be the vector subspace of $V$ generated by $M$; there exists a linear form $f$ on $M_0$ such that $f^{-1}(1) = M$; it is obvious that $f(x) \leq p(x)$ for every $x \in M_0$; by Theorem I.1 $f$ admits an extension $\widehat{f}$ to $V$ such that $\widehat{f}(x) \leq p(x)$ for every $x \in V$; let us prove that $H = \widehat{f}^{-1}(1)$ is convenient: $M \subset H$ since $\widehat{f}$ extends $f$ and $H \cap A = \emptyset$ (for each $x_0 \in H$ consider the segment $[0, x_0]$ and use the fact that $A$ is open and

$\widehat{f}(x_0) \leq p(x_0))$. Finally, if $H$ was not closed we would have, since it is an hyperplane, $\overline{H} = V$, thus $H \cap A \neq \emptyset$, a contradiction.

From the preceding result we deduce the following *Separation Theorem.*

## I.4 Theorem

Let $V$ be a t.v.s. not necessarily Hausdorff; if $A \subset V$ is a non-empty open convex set and if $B \subset V$ is a non-empty convex set such that $A \cap B = \emptyset$, there exists a closed hyperplane $H$ separating $A$ and $B$, namely there exist $f \in V'$ and $a \in \mathbb{R}$ such that $H = f^{-1}(a)$ and $f(A) \subset (-\infty, a[$ and $f(B) \subset [a, \infty)$.

Moreover, if $B$ is open there exits such a $H$ strictly separating $A$ and $B$, namely $f(A) \subset (-\infty, a[$ and $f(B) \subset ]a, \infty)$.

**Proof** (sketch)

Let $C = A - B$; this set is open, convex and $0 \notin C$; it suffices to apply the preceding theorem to $C$ with $M = 0$; there exists $f \in V'$ such that $f(x) - f(y) \neq 0$ for every $x \in A$ and $y \in B$, the result follows since $A$ is open and convex and $B$ is convex.

In the following theorem we deal with a Hausdorff l.c.s. (locally convex space).

## I.5 Theorem

Let $V$ be a Hausdorff l.c.s.. Let $A \subset V$ and $B \subset V$ be two non-empty convex sets such that $A$ is compact, $B$ closed and $A \cap B = \emptyset$. Then there exits a closed hyperplane strictly separating $A$ and $B$.

**Proof** (sketch)

Since $V$ is a Hausdorff l.c.s. it is easy to find two convex open sets $A_0$ and $B_0$ such that $A_0 \cap B_0 = \emptyset$ and $A \subset A_0$ and $B \subset B_0$; the result follows from the preceding theorem.

We are now in a position to state a classical result, which is rather a restatement of the preceding theorem. The topological dual of $V$ is denoted by $V'$ and the weak topology on $V$ is denoted by $\sigma(V, V')$.

## I.6 Theorem

Let $V$ be a Hausdorff l.c.s.; if $A \subset V$ is a closed convex set then $A$ is also closed for the topology $\sigma(V, V')$.

In particular, if $B$ is a Banach space a norm closed convex set $A \subset B$ is $\sigma(B, B')$ closed.

## I.7 Definition

Let $V$ be a Hausdorff l.c.s.; if $A \subset V$, we set

$$A^0 = \{f : f \in V', f(A) \subset [-1, \infty[\}$$

$A^0$ is said to be the polar set of $A$ taken in $V'$.

Accordinng to this definition, if $A$ is stable under the maps $x \mapsto \lambda x$ for every $\lambda > 0$, then $A^0$ is the set of all positive continuous forms on $A$.

The following result, known as the *Bipolar Theorem*, is a direct consequence of Theorem I.5.

## I.8 Theorem

Let $V$ be a Hausdorff l.c.s. and $A \subset V$; if $A^0$ is the polar set of $A$ taken in $V'$ and if $A^{00}$ is the polar set of $A^0$ taken in $V$ then $V^{00}$ is the closed convex hull of $A$ and 0.

## I.9 Notes and Remarks

We have tried to follow, roughly speaking, a chronological path; however, in the finite dimensional setting, some of these results were obtained previously.

We do not try to discuss thoroughly the history of the above theorems; we refer to the book by N. Dunford and J.T. Schwartz [22] (pages 86 and 460), published in 1957, which contains many historical references and a summary of many related results. However, let us indicate the opinions of S. Banach, F. Riesz and B. Sz.-Nagy:

In his book, published in 1932, S. Banach [6] mentioned a paper of F. Riesz [56], published in 1910 and a Note of E. Helly [29], published in 1912 concerning the case of $L^p$ spaces.

In their French-language book, written in 1948 and published in 1953, F. Riesz and B. Sz.-Nagy [57] wrote that the proof of Hahn-Banach's Theorem (I.2. above, given in I.1.) follows an idea contained in the Note of Helly, mentioned above; however this is not exactly the opinion of Dunford and Schwartz [22] (page 86). Moreover, F. Riesz and B. Sz.-Nagy mentioned also a paper of M. Riesz [58], published in 1922, where another version of Hahn-Banach's Theorem is proved in the framework of ordered vector spaces for the space of continuous functions on $\mathbb{R}$ in the setting of the moments problem.

In fact there are two aspects in this theory: the *extension* or *analytical* point of view and the *separation* or *geometrical* point of view. In 1953 N. Bourbaki (e.v.t. first edition) gave a presentation in which the analytical form is deduced from the separation theorem; this is also the presentation of H.H. Schaefer [62]. In 1965 N. Bourbaki (e.v.t. second edition) used the same presentation as in our book; this is also the case of the book by N. Dunford and J. T. Schwartz mentioned above.

Finally let us indicate a play on words: in the books by N. Dunford and J.T. Schwartz and in the book by H.H. Schaefer the Hahn-Banach Theorem and the Separation Theorem are *separated*, in II.3. and V.1. for the former, in II.3. and II.9. for the latter.

Let us now prove the three results mentioned in the Introduction (2-3-4), due to M.G. Krein, V.L. Klee and T. Ando; to help the reader, we recall the definition of *normal cones* (Introduction, Def. 1) which was given in 1940 by M.G. Krein.

## I.10 Definition

Let $X$ be a convex cone contained in a Banach space $B$. We say that $X$ is a *normal* cone if there exists a $C > 0$ such that for all $x, y \in X$ one has $\|x\| \leq C\|x + y\|$.

We say that $X$ is *generating* if $B = X - X$.

It is obvious that the closure of a normal cone is also normal.

Let us discuss the preceding definition of normal cones; $B$ can be viewed as an ordered vector space, with positive cone $X$, as it is indicated in the beginning of the Introduction; if $x, y \in B$ we denote by $[x, y]$ the order interval $\{z : z \in B, x \leq z \leq y\}$; if $A \subset B$ we denote by $[A]$ the union of all $[x, y]$ for all $x, y \in A$. It is clear that if $A$ is convex then $[A]$ is also convex, and that $[[A]] = [A]$.

## I.11 Proposition

Let $X$ be a convex cone contained in a Banach space $B$ whose closed unit ball is denoted by $B_1$. The following properties are equivalent:

1) $X$ is a normal cone.

2) $[B_1]$ is bounded.

3) The set $[X \cap B_1]$ is bounded.

4) If $(x_n) \subset X$ converges to 0 and if $0 \leq y_n \leq x_n$ then $y_n \to 0$.

5) There exists an equivalent norm $p$ on $B$ such that for all $x, y \in X$ one has $p(x) \leq p(x+y)$.

**Proof**

1) $\Rightarrow$ 2): if $x \leq y \leq z$ with $x, z \in B_1$ then $0 \leq y - x \leq z - x$, thus $\|y - x\| \leq 2C$ and the result follows.

2) $\Rightarrow$ 3) is obvious.

3) $\Rightarrow$ 4): by contradiction, otherwise there would exist two sequences $0 \leq y_n \leq x_n$ in $X$ and $\epsilon > 0$ with $x_n \to 0$ and $\|y_n\| \geq \epsilon$; it suffices to consider the sequence $y_n/\|x_n\|$ which is unbounded and to note that $0 \leq y_n/\|x_n\| \leq x_n/\|x_n\|$.

4) $\Rightarrow$ 1): by contradiction, otherwise there would exist two sequences $(x_n)$ and $(y_n)$ in $X$ with $\|x_n + y_n\| = 1$ and $\|x_n\| \to \infty$; one has $0 \leq x_n/\|x_n\| \leq (x_n + y_n)/\|x_n\|$ and the latter sequence converges to 0.

2) $\Rightarrow$ 5): since $[B_1]$ is symmetric, convex and bounded it suffices to take as $p$ the gauge of $[B_1]$.

5) $\Rightarrow$ 2) is obvious.

The following example shows that the condition $[X \cap B_1] = X \cap B_1$ does not imply $[B_1] = B_1$: let $B = \mathbb{R}^2$ equipped with the norm $\|x\| = |x_1| + |x_2|$ where $x = (x_1, x_2)$ is ordered by the cone $X = \{x : x_1 \geq 0, x_2 \geq 0\}$; one has $[X \cap B_1] = B_1$ but if $x = (-1, 0)$ and $y = (0, 1)$ it is clear that $[x, y]$ is not contained in $B_1$.

In Chapter V we shall see that norms with the property indicated in 5) of the above proposition are quite useful.

In 1940 M.G. Krein [33] proved the following basic theorem (Introduction, Thm. 2); we recall that $B'$ is the continuous dual of $B$ and that $X^0 \subset B'$ is the polar cone of $X$.

## I.12 Theorem

If $X \subset B$ is a normal cone then $X^0$ is a generating cone in $B'$.

**Proof**

Let $f \in B'$; for $x \in X$, we set $p_f(x) = \sup\{f(y) : 0 \le y \le x\}$; by Definition I.10 then $p_f$ is a superlinear form on $X$; namely for all $x, y \in X$ one has $p_f(x)+p_f(y) \le p_f(x+y)$; for every $x \in X$ and $\lambda \ge 0$ one has $p_f(\lambda x) = \lambda p_f(x)$; consequently in the product space $B \times \mathbb{R}$ the set

$$X_f = \{(x,t) : x \in X, 0 \le t \le p_f(x)\}$$

is a convex cone. Let us prove that $(0,1) \notin \overline{X}_f$: otherwise, there would exist two sequences $(x_n)$ and $(y_n)$ in $X$ with $0 \le y_n \le x_n$ and such that $x_n \to 0$ and $f(y_n) \to 1$; this is a contradiction by Proposition I.11. 4), since $X$ is normal.

By Theorem I.5 there exists a closed hyperplane $H$ strictly separating $X_f$ and $(0,1)$; thus there exist $g \in B'$ and $\alpha, \beta \in \mathbb{R}$ such that

$$H = \{(x,t) : g(x) + \alpha t = \beta\}$$

with $\alpha > \beta$ and for all $0 \le y \le x$: $g(x) + \alpha f(y) \le \beta$.

The last inequality (with $x = y = 0$) implies $\beta \ge 0$, thus $\alpha > 0$;

by homogeneity $g(x) + \alpha f(y) \le 0$ for $0 \le y \le x$, thus:

with $y = 0$ we get $g(x) \le 0$; in other words $g \in -X^0$,

with $y = x$ we get $g(x) + \alpha f(x) \le 0$; in other words $g + \alpha f \in -X^0$; the result follows since $\alpha \ne 0$.

In the same paper M.G. Krein proved the following result; $X^0$ denotes the polar cone of $X$ taken in $B'$.

## I.13 Theorem

If $X \subset B$ is a normal cone there exists a $C > 0$ such that every $f \in B'$ admits a decomposition $f = f_1 - f_2$ where $f_1, f_2 \in X^0$ and $\|f_1\| + \|f_2\| \le C\|f\|$.

**Proof**

The set $\Gamma = \{f : f \in X^0, \|f\| \le 1\}$ is $\sigma(B', B)$ compact and this is also the case for $\Gamma - \Gamma$, thus this set is norm closed; since, by the preceding theorem, $B' = \cup\, n(\Gamma - \Gamma)$ it follows from Baire's Theorem that $(\Gamma - \Gamma)$ contains an open ball of $B'$ and it is easy to finish the proof.

In 1955 V.L. Klee [31] proved the following lemma, interesting in its own-right.

## I.14 Lemma

Let $X \subset B$ be a closed convex cone (not necessarily proper) and $X_1 = X \cap B_1$; the vector space $E = X - X$, equipped with the gauge of $X_1 - X_1$ is a Banach space.

**Proof**

Let $p$ be the gauge of $X_1 - X_1$; it is obvious that $p$ is a norm on $E$. Let $(x_n)$ be a sequence in $E$ with $p(x_n) \leq 2^{-n}$; there exist two sequences $(y_n)$ and $(z_n)$ in $X$ such that

$$x_n = y_n - z_n \text{ with } \|y_n\| \leq 2^{-n} \text{ and } \|z_n\| \leq 2^{-n}$$

Since $X$ is closed the elements $y_0 = \sum_1^\infty y_n$ and $z_0 = \sum_1^\infty z_n$ exist in $X$; let us prove that $\sum_1^n x_k \to y_0 - z_0$ for the norm $p$.

It suffices to prove that $\sum_1^\infty y_k - \sum_1^n y_k = \sum_{n+1}^\infty y_k \to 0$ for the norm $p$; one has $\| \sum_{n+1}^\infty y_k \| \leq \sum_{n+1}^\infty 2^{-k} \to 0$; the result follows by the way things were set up.

Here is the theorem of V.L. Klee (Introduction, Thm. 3), proved in 1955:

## I.15 Theorem

If $X \subset B$ is a closed convex generating cone (not necessarily proper) then there exists a $C > 0$ such that every $x \in B$ admits a decomposition $x = x_1 - x_2$ with $x_1, x_2 \in X$ and $\|x_1\| + \|x_2\| \leq C\|x\|$.

Moreover, every linear form $f$ on $B$ such that $f(x) \geq 0$ for every $x \in X$ is continuous.

**Proof**

With the notations of the preceding lemma, the canonical map from $B = X - X$, equipped with the norm $p$, onto $B$, equipped with its original norm, is continuous; by a classical theorem due to Banach (open mapping theorem) the two norms are equivalent.

If $f$ was not continuous there would exist a sequence $(x_n)$ in $X_1$ with $f(x_n) \geq 4^n$; let $x_0 = \sum x_n/2^n$; considering $f(x_0)$ it is easy to get a contradiction, since $f$ is assumed to be positive on $X$.

## I.16 Remark

It follows from the preceding theorem that (under the hypothesis of I.15) the polar cone $X^0$ is *weakly complete* for the duality with $B$, namely $X^0$ is complete when $B'$ is equipped with the *uniform structure* associated to the family of pseudo-metrics $d_x(f_1, f_2) = |f_1(x) - f_2(x)|$ where $x \in B$. Indeed, any limit of positive linear forms (on $X$) is positive (on $X$). This notion was introduced by G. Choquet in the framework of Integral Representation Theory (see the Posface of G. Choquet in [14]).

In 1962 T. Ando [4] proved the following theorem (Introduction, Thm. 4):

## I.17 Theorem

Let $X \subset B$ be a closed convex cone (not necessarily proper); if $X^0$ is a normal cone in $B'$, then $X$ is generating in $B$.

**Proof**

The polar set of $X^0 + B'_1$ is $X_1 = X \cap B_1$; moreover the convex set $X^0 + B'_1$ is closed for $\sigma(B', B)$ because $X^0$ is closed for this topology and $B'_1$ is compact for this topology. Consequently the polar set of $X_1 \cup -X_1$ is the convex set $(X^0 + B'_1) \cap (-X^0 + B'_1)$ which is closed for $\sigma(B', B)$; moreover this set is bounded since $X^0$ is a normal cone; by the Bipolar Theorem I.8 it follows that the closed convex hull of $X_1 \cup -X_1$ contains $\alpha B_1$ for some $\alpha > 0$.

Let us resume the notations of Lemma 1.14: let $\varphi$ be the canonical map from $E$ into $B$; by the preceding paragrah $\overline{\varphi(E_1)}$ contains $\alpha B_1$. By a classical theorem $\varphi$ is onto and the result follows from the open mapping theorem.

## I.18 Notes and remarks

Let us explain why normal cones are so important, in particular in the framework of *Spectral Theory* concerning positive operators.

To begin with, we recall (without proof) a classical theorem due to Pringsheim:

If $f(z) = \sum_0^\infty a_n z^n$ is an analytic function with $a_n \geq 0$ for every $n$, and if the radius of convergence $r$ of $f$ satisfies $0 < r < \infty$, then $f$ is singular at $z = r$, namely $f$ does not extend to an analytic function on a neighbourhood of $r$.

The same result holds if $f$ takes its values in a H.l.c.s. $E$, equipped with a closed convex cone $X$ such that $E' = X^0 - X^0$, whenever $a_n \in X$ for every $n$. It is the case if $E$ is a Banach space and $X \subset E$ is a closed normal cone.

It is easy to see that a real Banach space can be isometrically embedded in a *complex* Banach space, so that one can use Spectral Theory:

Recall that, if $E$ is a complex Banach space and if $u : E \to E$ is continuous, the *spectrum* $\sigma(u)$ of $u$ is the set of all $\lambda \in \mathbb{C}$ such that $(\lambda I - u)^{-1}$ does not exist; it is well known that (if $E \neq 0$) $\sigma(u)$ is a non-empty compact subset of $\mathbb{C}$; we denote by $r(u)$ the *spectral radius* of $u$, namely the supremum of $|\lambda|$ for all $\lambda \in \sigma(u)$.

If $\rho(u) = \mathbb{C} \setminus \sigma(u)$, it is well known that the map $\lambda \mapsto R(\lambda) = (\lambda I - u)^{-1}$ is analytic in $\rho(u)$; this map is called the *resolvent* of $u$.

One has the following theorem:

Let $B$ be a (real) Banach space ordered by a closed generating normal cone $X$. If $u : B \to B$ is positive ($u(X) \subset X$) then the spectral radius $r(u)$ belongs to the spectrum of $u$, namely $(u - r(u)I)^{-1}$ does not exist.

If we assume only that $X$ is a closed proper convex cone and that $B = \overline{X - X}$ this result remains true whenever the resolvent map $\lambda \mapsto R(\lambda)$ has a pole on the spectral circle $\{z : |z| = r(u)\}$; in this case $r(u)$ is a pole of the resolvent of maximal order on the spectral circle.

The proof consists in using Pringsheim's Theorem with $f$ the map $z \mapsto R(1/z)$.

The following theorem is due to Krein and Rutman:

If $X \subset B$ (real Banach space) is a closed convex cone and satisfies $B = \overline{X - X}$, then for every positive compact operator $u : B \to B$ with $r(u) > 0$, $r(u)$ is a pole of the resolvent of $u$, of maximal order on the spectral circle $\{z : |z| = r(u)\}$, and $u$ admits an eigenvector in $X$. The same result holds for $u' : B' \to B'$.

For more details we refer to the Appendix of H.H. Schaefer's book [62].

# Chapter II

# Theorems of factorization for operators from a Banach space to a space $L^1$

The results of this Chapter, mainly due to B. Maurey [44], will be useful in Chapter IV. We first deal with the following problem (Theorem II.7) : let $E$ be a Banach space and $u$ be a continuous linear operator from $E$ to $L^1(\Omega, \mathcal{A}, m)$, where $(\Omega, \mathcal{A}, m)$ is a measure space. Let $1 < p < \infty$; we want to find necessary and sufficient conditions ensuring that $u$ *factors* through $L^p(\Omega, \mathcal{A}, m)$ in the following sense:

## II.1 Definition

We say that $u : E \to L^1(\Omega, \mathcal{A}, m)$ *factors* through $L^p(\Omega, \mathcal{A}, m)$ if there exist $g \in L^{p'}(\Omega, \mathcal{A}, m)$ $(1/p + 1/p' = 1)$, and some constant $C > 0$ such that, for every $x \in E$, one has $\{u(x) \neq 0\} \subset \{g \neq 0\}$ (almost everywhere) and

$$\|u(x)/g\|_p \leq C\|x\|$$

(with $0/0 = 0$).

The minimal constant $C$ for $\|g\|_{p'} \leq 1$ is denoted by $C_{1,p}(u)$.

When the conditions of the above definition are satisfied $u$ admits obviously the factorization $x \mapsto m_g(u(x)/g)$, where $m_g : L^p(\Omega, \mathcal{A}, m) \to L^1(\Omega, \mathcal{A}, m)$ is defined by $f \mapsto gf$.

Then we prove an extension, due to B. Maurey [44], of a lemma of Rosenthal (Lemma II.8) which is, roughly speaking, a deep investigation of what happen when $u$ does not satisfy the conditions of the preceding definition.

The following two lemmas will be useful to prove Lemma II.4, which is classical in Convex Analysis, and will enable us to prove Theorem 7.

## II.2 Lemma

Let $V$ be a Hausdorff l.c.s. and $K \subset V$ be a non-empty closed convex set. Let $f : K \rightarrow ]-\infty, \infty]$ be convex, lower semicontinuous (for the topology induced on $K$ by $V$); then $f$ is the supremum of the set of all the continuous affine functions which are majorized by $f$ on $K$.

Consequently, $f$ is also lower semicontinuous for the weak topology $\sigma(V, V')$.

**Proof**

If $f$ is identical to $+\infty$ the lemma is obvious.

Otherwise, in the space $V \times \mathbb{R}$ let $\Gamma$ be the epigraph of $f$, namely:

$$\Gamma = \{(x,t) : x \in K,\ t \geq f(x)\}$$

$\Gamma$ is closed in $V \times \mathbb{R}$ since $f$ is assumed to be lower semicontinuous; we apply the Separation Theorem I.4 to the closed convex set $\Gamma$ and to each point $(x_0, t_0)$ with $x_0 \in K$ and $t_0 < f(x_0)$

Let $H$ be a closed hyperplane in $V \times \mathbb{R}$ strictly separating $\Gamma$ and $(x_0, t_0)$; there exist $\varphi \in V'$ and $\alpha, \beta \in \mathbb{R}$ such that:

$$H = \{(x,t) : x \in V,\ t \in \mathbb{R},\ \varphi(x) + \alpha t + \beta = 0\}$$

$$\varphi(x) + \alpha(f(x) + r) + \beta < 0 \text{ for every } x \in K \text{ with } f(x) < \infty \text{ and every } r \geq 0$$

$$\varphi(x_0) + \alpha t_0 + \beta > 0$$

By the first inequality $\alpha \leq 0$ and, using the second inequality, it is easy to see that $\alpha < 0$ if $f(x_0) < \infty$, thus, in this case, we can assume that $\alpha = -1$.

Consequently, the map $x \mapsto \varphi(x) + \beta$ satisfies $\varphi(x) + \beta < f(x)$ for every $x \in K$ and

$\varphi(x_0) + \beta > t_0$.

The result follows if $f$ is finite at every point, because the upper envelope of a family of continuous functions is lower semicontinuous.

If $f(x_0) = +\infty$, since $f$ is not indentical to $+\infty$, by the first part of the proof we can assume that $f \geq 0$ on $K$; let us resume the above notations: we have seen that $\alpha \leq 0$; if $\alpha < 0$ we get the result as above; if $\alpha = 0$ then:

$$\varphi(x) + \beta < 0 \text{ for every } x \in K \text{ with } f(x) < +\infty$$

$$\varphi(x_0) + \beta > 0$$

For every $\alpha_0 < 0$ the inequality

$$\varphi(x) + \alpha_0(f(x) + r) + \beta < 0$$

holds for every $x \in K$ with $f(x) < +\infty$ and all $r \geq 0$, since $f \geq 0$ on $K$; since the inequality

$$\varphi(x_0) + \alpha_0 t_0 + \beta > 0$$

holds if $|\alpha_0|$ is small enough we can finish the proof as above.

## II.3 Lemma

Let $V$ be a Hausdorff l.c.s. and $K \subset V$ be a non-empty compact convex set; if $\mu$ is a probability Radon measure on $K$ there exists a unique $x \in K$, called the *barycenter* of $\mu$ and denoted by $r(\mu)$, such that, for every continuous affine function $f$ defined on $K$:

$$\mu(f) = f(x)$$

**Proof**

For every $x \in K$ let $\epsilon_x$ be the Dirac measure at $x$ (namely the measure $f \mapsto f(x)$ for every $f \in \mathcal{C}(K)$). It is a well-known fact that $\mu$ is the limit of an ultrafilter $\mathcal{U}$ on the set of all finite sums $\sum \alpha_x \epsilon_x$, where $x \in K$ and $\alpha_x > 0$ with $\sum \alpha_x = 1$, for the duality with the space $\mathcal{C}(K)$ of all continuous functions on $K$.

To every finite sum $s = \sum \alpha_x \epsilon_x$ we associate the element $x_s = \sum \alpha_x x$; since $\alpha_x > 0$ and $\sum \alpha_x = 1$ one has $x_s \in K$; the element $\lim_{\mathcal{U}}(x_s)$ is convenient because for every continuous affine function $f$ on $K$ one has $f(x_s) = f(\sum \alpha_x \epsilon_x)$ and it suffices to take the limit along $\mathcal{U}$ to get the result. This element is unique because $V$ is Hausdorff.

The following lemma is classical in Convex Analysis.

## II.4 Lemma

Let $V$ be a Hausdorff l.c.s. and $K \subset V$ be a non-empty compact convex set. Let $\Gamma$ be a non-empty convex set of upper semicontinuous concave functions $f : K \to [-\infty, \infty[$; suppose that for each $f \in \Gamma$ there exists $x \in K$ such that $f(x) \geq 0$. Then there exists $x_0 \in \Gamma$ such that $f(x_0) \geq 0$ for every $f \in \Gamma$.

**Proof**

Let $\Gamma_a$ be the set of all continuous affine functions $\varphi : K \to \mathbb{R}$ such that there exists $f \in \Gamma$ (depending on $\varphi$) satisfying $\varphi \geq f$ on $K$. It is obvious that

$\Gamma_a$ is non-empty, since every element of $\Gamma$ is bounded, and that $\Gamma_\alpha$ is a convex set. Let

$$C_- = \{g : g \in \mathcal{C}(X), g(x) < 0, \forall x \in K\}$$

By hypothesis $\Gamma_a$ and $C_-$ are disjoint; since $\mathcal{C}_-$ is open, by Theorem I.4 there exist a Radon measure $\mu \neq 0$ on $K$ and $a \in \mathbb{R}$ such that:

$$\mu(g) \leq a \text{ for every } g \in \mathcal{C}_- \text{ and } \mu(\varphi) \geq a \text{ for every } \varphi \in \Gamma_a$$

Since $\mathcal{C}_-$ is a cone we have $\mu(g) \leq 0$ for every $g \in \mathcal{C}_-$; in other words $\mu \geq 0$; it follows that $a \geq 0$ because $\mu(-1/n) = -\mu(1)/n$ for each integer $n$; we can thus assume that $\mu$ is a probability Radon measure.

Let us prove that $r(\mu)$ (defined in the preceding lemma) is convenient:

one has $\varphi(r(\mu)) = \mu(\varphi) \geq 0$ for every $\varphi \in \Gamma_a$ and it suffices now to apply Lemma II.2 to the elements of $\Gamma$.

## II.5 Lemma

Let $(\Omega, \mathcal{A}, m)$ be a measure space, $f \in L^1(m)$ and $1 < p < \infty$. Let

$$K = \{g : \|g\|_{p'} \leq 1, \ g \geq 0\}$$

If $f \geq 0$, the map $\Psi : K \to [0, \infty]$ defined by $g \mapsto \int (f^p/g^p)dm$ (with $0/0 = 0$) is convex, lower semicontinuous for the weak topology and not identically infinite.

**Proof**

The convexity of $\Psi$ follows from the fact that $t \mapsto 1/t^p$ is convex on $]0, +\infty[$; by Lemma II.2 it suffices to prove that $\Psi$ is lower semicontinuous for the norm topology: if $g_n \to g_0$ in $L^1(m)$ then there exists a subsequence, denoted again $g_n$ for short, such that $g_n \to g_0$ almost everywhere; it suffices to apply Fatou's Lemma to get $\Psi(g_0) \leq \liminf \Psi(g_n)$.

Let $g = f^{1/p'}$ if $f(x) > 0$ and $g = 0$ otherwise, then $g \in L^{p'}$ and $\Psi(g) = \int f dm < \infty$.

The theorem we are aiming for is a consequence of the following more general result.

## II.6 Theorem

Let $(\Omega, \mathcal{A}, m)$ be a measure space. Let $(f_i)$, $i \in I$, be a family of functions in $L^1(\Omega, \mathcal{A}, m)$. Then, for $1 < p < \infty$, the following two properties are equivalent:

1) For every $(\alpha_i) \in \mathbb{R}^{(I)}$ one has:

$$\int (\sum_{i \in I} |\alpha_i||f_i|^p)^{1/p} dm \leq (\sum_{i \in I} |\alpha_i|)^{1/p}$$

2) There exists $g \in L^{p'}(\Omega, \mathcal{A}, m)$ with $\|g\|_{p'} \leq 1$ such that for every $i \in I$ one has:

$$\{f_i \neq 0\} \subset \{g \neq 0\} \text{ (almost everywhere) and} \|f_i/g\|_p \leq 1$$

(with $0/0 = 0$)

**Proof**

1) $\Rightarrow$ 2): we can assume that $f_i \geq 0$ for every $i \in I$; for $\alpha = (\alpha_i) \in (\mathbb{R}^+)^{(I)}$, with the notations of lemma II.5, let $\Psi_\alpha$ be the function on $K$ associated to $(\sum \alpha_i f_i^p)^{1/p}$, namely:

$$\Psi_\alpha(g) = \int \frac{(\sum \alpha_i f_i^p)}{g^p} dm$$

Let $\Gamma$ be the set of all the maps from $K$ to $[-\infty, \infty[$ of the form $(\sum \alpha_i - \Psi_\alpha)$; it is obvious that $\Gamma$ is a convex cone consisting of upper semicontinuous functions on $K$; let us check that, for every $\alpha = (\alpha_i) \in (\mathbb{R}^+)^{(I)}$, there exists $g_\alpha \in K$ with $\sum \alpha_i - \Psi_\alpha(g_\alpha) \geq 0$:

If $f_\alpha = (\sum \alpha_i f_i^p)^{1/p}$, let $g_\alpha = f_\alpha^{1/p'}/(\sum \alpha_i)^{1/pp'}$; then:

$$\|g_\alpha\|_{p'}^{p'} = (\int f_\alpha dm)/(\sum \alpha_i)^{1/p} \leq 1$$

and

$$\Psi_\alpha(g_\alpha) = (\int \frac{f_\alpha^p}{f_\alpha^{p/p'}} dm)(\sum \alpha_i)^{1/p'} = (\int f_\alpha dm)(\sum \alpha_i)^{1/p'} \leq (\sum \alpha_i)$$

since $\int f_\alpha dm \leq (\sum |\alpha_i|)^{1/p}$ by condition 1).

By Lemma II.4 applied to $\Gamma$, there exists $g \in K$ such that $\sum \alpha_i - \Psi_\alpha(g) \geq 0$ for every $\alpha \in (\mathbb{R}^+)^{(I)}$; the result follows.

2) $\Rightarrow$ 1): by Holder's Inequality we have

$$\int (\sum \alpha_i f_i^p)^{1/p} dm \leq \left(\int \frac{(\sum \alpha_i f_i^p)}{g^p} dm\right)^{1/p} \|g\|_{p'} \leq (\sum \alpha_i)^{1/p}$$

and the lemma follows.

The following result answers the question raised in the introductory paragraph of this Chapter.

# II.7 Theorem

Let $(\Omega, \mathcal{A}, m)$ be a measure space. Let $E$ be a Banach space and $u : E \to L^1(\Omega, \mathcal{A}, m)$ be a continuous linear operator. Then, for every $1 < p < \infty$, the following two properties are equivalent:

1) There exists a constant $C_1 > 0$ such that for every finite sequence $x_1,,,x_n$ in $E$ one has:

$$\int (\sum_1^n |u(x_i)|^p)^{1/p} dm \leq C_1 (\sum_1^n \|x_i\|^p)^{1/p}$$

2) There exist $g \in L^{p'}(\Omega, \mathcal{A}, m)$ and a constant $C_2 > 0$ such that one has:

$$\{u(x) \neq 0\} \subset \{g \neq 0\} \text{ a.e. and } \|u(x)/g\|_p \leq C_2 \|x\| \text{ for every } x \in E.$$

(with $0/0 = 0$)

Moreover, when these conditions are satisfied, the minimal constant $C_1$, denoted by $C_{1,p}(u)$, is nothing but:

$$C_{1,p}(u) = \inf\{C_2 : \text{ such that 2) holds with } \|g\|_{p'} = 1\}$$

**Proof**

We apply the preceding theorem to the set $\{u(x) : x \in E, \|x\| = 1\}$.

We now deal with *Rosenthal's Lemma*

# II.8 Lemma (Rosenthal's Lemma)

For every integer $n$, every $1 < p < \infty$ and every $0 < \delta < 1$, there exists a number $N = N(n, p, \delta)$ such that:

If $u$ is a norm one operator from a Banach space $E$ to a space $L^1(\Omega, \mathcal{A}, m)$, admitting a factorization through $L^p(\Omega, \mathcal{A}, m)$ satisfying $C_{1,p}(u) \geq N$, there exists $g \in L^{p'}(\Omega, \mathcal{A}, m)$ with $\|g\|_{p'} = 1$ and $g \geq 0$, such that if we denote by $P$ the probability $g^{p'} dm$ one has:

1) $\{u(x) \neq 0\} \subset \{g \neq 0\}$ almost everywhere for $x \in E$, with respect to $m$.

2) If $u_1$ denotes the map $x \mapsto u(x)/g^{p'}$ from $E$ to $L^1(\Omega, \mathcal{A}, P)$, then:

$$\int |u_1(x)|^p dP \leq (C_{1,p}(u) + 1)^p \|x\|^p \text{ for every } x \in E.$$

3) There exist $n$ pairwise disjoint subsets $A_1,,,A_n$ of $\Omega$ and norm one elements $x_1,,,x_n$ of $E$ such that for every $1 \leq i \leq n$ one has:

$$\int_{A_i} |u_1(x_i)|^p dP \geq \delta^p (C_{1,p}(u))^p$$

**Proof**

By definition of $C_{1,p}(u)$, there exists $g \in L^{p'}(\Omega, \mathcal{A}, m)$, with $\|g\|_{p'} = 1$ and $g \geq 0$, such that condition 1) is satisfied and $\|u(x)/g\|_p \leq (C_{1,p}(u)+1)\|x\|$ for every $x \in E$.

Condition 2) is satisfied by a simple calculation.

One has $\|u_1\| = \|u\| = 1$; to prove that $C_{1,p}(u_1) = C_{1,p}(u)$, it suffices to check that for every finite sequence $x_1,,,x_n$ in E one has:

$$\int (\sum_1^n |u(x_i)|^p)^{1/p} dm = \int (\sum_1^n |u_1(x_i)|^p)^{1/p} dP$$

and to apply Theorem II.7 ; to simplify notations we write $C_{1,p}$.

It remains to prove 3); first, note that, for every $x \in E$ with $\|x\| = 1$ and for every $K > 0$, if $A = \{|u_1(x)| > K\}$, one has $P(A) \leq 1/K$ and $\int_{A^c} |u_1(x)|^p \leq K^{p-1}$ since $\|u_1\| = 1$; by definition of $C_{1,p}$ there exists $x_0 \in E$, $\|x_0\| = 1$, such that $\int |u_1(x_0)|^p \geq (C_{1,p} - 1)^p$; consequently, if $A_0 = \{|u_1(x_0)| > K\}$ one has:

$$\int_{A_0^c} |u_1(x_0)|^p dP \geq (C_{1,p} - 1)^p - K^{p-1}$$

The preceding considerations make it natural the following lemma which will imply the result we are aiming for.

## II.9 Lemma (with the hypothesis of the preceding lemma)

There exist $K$ and $N = N(p, n, \delta)$ such that, if $C_{1,p}(u) \geq N$, for every measurable set $A \subset \Omega$ with $P(A) \leq n/K$, there exist $x_A \in E, \|x_A\| = 1$, and a measurable set $B \subset A^c$ with $P(B) \leq 1/K$, satisfying:

$$\int_B |u_1(x_A)|^p dP \geq \delta^p (C_{1,p}(u))^p$$

**Proof**

Let $g_a = a^{1/p'} 1(A) + b^{1/p'} 1(A^c)$ be a norm one element of $L^{p'}(P)$; one has $aP(A) + b(1 - P(A)) = 1$.

Notice that we can choose $a$ arbitrary: indeed it suffices to take $K$ so big that $na/K < 1$; moreover it follows that $b \geq 1 - aP(A) \geq 1 - na/K$.

We shall arrange that $a$ be "great" and that $b$ be close to 1; to this end we shall take $na/K$ "small".

By definition of $C_{1,p}$ there exists a norm one element $x \in E$ such that

$$\int |u_1(x)/g_a|^p dP \geq (C_{1,p} - 1)^p$$

We set $B = A^c \cap \{|u_1(x)| > K\}$ and $C = \Omega \setminus A \cup B$; one has $P(B) \leq 1/K$ and:

$$\int_A |u_1(x)/g_a|^p dP \leq (\int |u_1(x)|^p dP)/a^{p/p'} \leq (C_{1,p} + 1)^p/a^{p/p'}$$

$$\int_C |u_1(x)/g_a|^p dP \leq K^{p-1}/b^{p/p'}$$

Consequently one has:

$$(\int_B |u_1(x)|^p dP)/b^{p/p'} \geq (C_{1,p} - 1)^p - (C_{1,p} + 1)^p/a^{p/p'} - K^{p-1}/b^{p/p'}$$

Now it is easy to finish the proof:

We choose $a$ such that $1 - 1/a^{p/p'} > \delta$, and then $K$ such that $b^{p/p'}(1 - 1/a^{p/p'}) > \delta$ (note that $b \geq 1 - na/K$); now it suffices to choose $C_{1,p}$ great enough to satisfy:

$$\int_B |u_1(x)|^p dP \geq \delta^p C_{1,p}^p$$

Lemma II.9 is proved; we can finish the proof of Lemma II.8:

We apply Lemma II.9 with $A = \emptyset$; we denote by $x_1$ the norm 1 element of $E$ and by $A_1$ the set $B$ given by this lemma; we apply again the lemma with $A_1$ and so on.

## II.10 Notes and Remarks

1) An origin of the factorization theory described in this Chapter can be found in a paper by E.M. Nikishin [51], published in 1970, who proved the following result:

   Let $L^0(\Omega, \mathcal{A}, P)$ be the space of all (almost everywhere) finite measurable functions defined on a measurable space $(\Omega, \mathcal{A}, P)$, where $P$ is a probability, equipped with the topology of convergence in measure, and let $E$ be a Banach space. For every continuous linear operator $u : E \to L^0(P)$, for every $0 < p < 1$ and every $\epsilon > 0$, there exist a measurable set $\Omega_\epsilon \subset \Omega$ with $P(\Omega \setminus \Omega_\epsilon) \leq \epsilon$, and a constant $C$ such that:

$$\int_{\Omega_\epsilon} |u(x)|^p \leq C\|x\|^p \quad \text{for every } x \in E.$$

This theorem implies that, for each $0 < p < 1$, there exist a continuous linear operator $u_p : E \to L^p(P)$ and $g_p \in L^0(P)$ such that $u$ admits the factorization $u = M_{g_p} \circ u_p$, where $M_{g_p}$ is the multiplication by $g_p$.

2) Note that, in 1956, A. Grothendieck [26], [27] proved that if $E$ is a $L^\infty$ space the conditions of Theorem II.7 are always satisfied for $p = 2$ for all operators $u$.

3) In 1973 H.P. Rosenthal [59] studied the subspaces of a space $L^1(P)$, where $P$ is a probability; with the notations of Rosenthal's Article, if $R$ is a closed subspace of $L^1(P)$, Theorem 1 of his Article is nothing but our Theorem II.6 with $E = R$ and where $u$ is the canonical injection $i$ from $R$ into $L^1(P)$. In fact, the point view of H.P. Rosenthal is a dual point of view: it consists to study the canonical quotient map $i'$ from $L^\infty(P)$ onto $R'$: for every $1 < p < \infty$ Rosenthal proved that $i$ factors through $L^p(P)$ iff the transposed map $i' : L^\infty(P) \to R'$ is $p'$-summing.

Lemma II.8 (Rosenthal's Lemma) is proved in Lemma 6 of Rosenthal's Article in the particular case indicated above of a subspace $R$ of $L^1(P)$ for the canonical injection $i$; in fact Lemma II.8 is the sublemma (and the main point) of Lemma 6 of Rosenthal's Article; we have followed his proof.

B. Maurey [44] extended this lemma in the setting described in Lemma II.8.

4) In fact B. Maurey [44] proved a more general version of Theorem II.7:

Let $1 < q < p < \infty$ and $1/q = 1/p + 1/r$, for every Banach space $E$ and every operator $u : E \to L^q(\Omega, \mathcal{A}, m)$ the following two properties are equivalent:

a) There exists a constant $C_a$ such that, for every finite sequence $x_1, , , x_n$ in $E$, one has:

$$\|(\sum_1^n |u(x_i)|^p)^{1/p}\|_q \leq C_a(\sum_1^n \|x_i\|^p)^{1/p}$$

b) There exist $g \in L^r(\Omega, \mathcal{A}, m)$ and a constant $C_b$ such that, for every $x \in E$, one has:

$$\{u(x) \neq 0\} \subset \{g \neq 0\} \text{ and } \|u(x)/g\|_p \leq C_b\|x\|$$

(with $0/0 = 0$)

Moreover, when these conditions are satisfied the minimal constant $C_a$ is equal to the infimum of the constants $C_b$ with $\|g\|_r = 1$. Let $C_{q,p}(u)$ denote this number.

The proof is similar to that of Theorem II.7 but we only use here the case $q = 1$.

5) Maurey's Theory was the starting point of results of J.L. Krivine [34] concerning Banach lattices: to begin with, note that if $V$ is a Banach lattice and if $1 \leq p \leq \infty$ for every finite sequence $v_1,,,v_n$ in $V$, one can define an element denoted by $(\sum_1^n |v_i|^p)^{1/p}$ as the supremum of all the elements $\sum_1^n \alpha_i v_i$, where $(\sum_1^n |\alpha_i|^{p'})^{1/p'} = 1$; this is immediate if $V$ is a Banach lattice of functions on some set. This element is called the "Krivine element".

Inspired by condition a) of 4) above J.L. Krivine [34] introduced the following two definitions:

a) If $E$ is a Banach space, $V$ a Banach lattice and if $1 \leq p \leq \infty$, an operator $u : E \to V$ is said to be of *type* $\geq p$ if there exists some constant $K$ such that for every $n$ and every finite sequence $x_1,,,x_n$ in $E$ one has:

$$\|(\sum_1^n |u(x_i)|^p)^{1/p}\| \leq K(\sum_1^n \|x_i\|^p)^{1/p}$$

Let $K^{(p)}(u)$ be the minimal constant such that the preceding inequality holds.

In the modern terminology $u$ is said to be *p-convex*. Note that J.L. Krivine did not define the type of $u$, but that his terminology makes sense since one can prove that if $u$ is of type $\geq p$ then $u$ is also of type $\geq p_1$ for every $1 \leq p_1 < p$. One has $K^{(1)}(u) = \|u\|$.

It is immediate that this definition is the same as condition a) of 4) above when $V = L^q(\Omega, \mathcal{A}, m)$.

b) If $E$ is a Banach space, $V$ a Banach lattice and if $1 \leq p \leq \infty$, an operator $v : V \to E$ is said to be of *type* $\leq p$ if there exists some constant $K$ such that for every $n$ and every finite sequence $x_1,,,x_n$ in $V$ one has:

$$(\sum_1^n \|v(x_i)\|^p)^{1/p} \leq K\|(\sum_1^n |x_i|^p)^{1/p}\|$$

Let $K_{(p)}(u)$ be the minimal constant such that the preceding inequality holds.

In the modern terminology $v$ is said to be *p-concave.* Note that J.L. Krivine did not define the type of $v$, but again his terminology makes sense since one can prove that if $v$ is of type $\leq p$ then $v$ is also of type $\leq p_1$ for every $p < p_1 \leq \infty$. One has $K_{(\infty)}(v) = \|v\|$.

c) From these notions J.L. Krivine [34] was abble to derive a new proof of Grothendieck's Inequality stated in the following form (see also VIII.15.(2)).

Let $L$ and $M$ be two Banach lattices; then, for every continuous linear map $U : L \longrightarrow M$, for every finite sequence $x_1,,,x_n$ in $L$ one has:

$$\|(\sum_1^n |U(x_i)|^2)^{1/2}\| \leq K_G\|U\|\|(\sum_1^n |x_i|^2)^{1/2}\|$$

where $K_G$ is Grothendieck's Constant.

# Chapter III

# Conical Measures and weakly complete convex cones

In this Chapter we prove some results concerning *conical measures* and *weakly complete* convex cones which will be useful throughout our book. The results are stated in the general setting of Hausdorff locally convex spaces.

Conical measures were introduced in 1962 by G. Choquet in the framework of his Theory of Integral Representation; we refer to [14] where this Theory is developped. In the present book, conical measures will be a quite useful tool, but for a purpose which has nothing to do with the Theory of Integral Representation: taking account of Theorem II.7, Lemmas III.14, 15 will explain perfectly the key role of conical measures.

In this Chapter, $E$ denotes a Hausdorff l.c.s. with topological dual $E'$ and algebraic dual $E^*$.

## III.1 Definitions

Let $h(E)$ be the *vector lattice* of functions on $E$ generated by the elements of $E'$; a *positive conical measure* $\mu$ on $E$, for the duality with $E'$, is a positive linear form on the space $h(E)$; we denote by $M^+(E)$ the convex cone of those $\mu$.

We set

$$M(E) = M^+(E) - M^+(E)$$

Notice that, since $h(E)$ is a vector lattice, it follows that $M(E)$ is a *Dedekind complete vector lattice*, namely, if $\mu_\alpha$ is a majorized upward directed family in $M^+(E)$ then $\lim \mu_\alpha$ exists in $M^+(E)$.

Here are some examples of conical measures: if $K \subset E$ is a compact set and if $m$ is a Radon measure on $K$ then the map $f \mapsto m(f)$ is obviously a conical measure on $E$; note that if $x \in E$ and $k > 0$ then $\epsilon_x = k\epsilon_y$ with $y = x/k$ as conical measures. The question of the representation of conical measures will be tackled in Chapter VI.

## III.2 Proposition

Every $f \in h(E)$ can be written $f = \sup(\ell_1,,,\ell_n) - \sup(\ell_{n+1},,,\ell_{n+m})$, where $\ell_1,,,\ell_{n+m} \in E'$, and there exist a finite family of convex cones $X_1,,,X_k$ in $E$, with $E \subset \bigcup_1^k X_i$, and $\varphi_1,,,\varphi_k \in E'$ such that, for every $1 \leq i \leq k$, one has $f = \varphi_i$ on $X_i$.

**Proof**

The first part of the proposition is easy and left to the reader. If $f = \sup(\ell_1,,,\ell_n)$, for every $1 \leq i \leq n$, we set: $X_i = \{x : x \in E,\ \ell_i(x) = f(x)\}$; it is easy to see that $X_i$ is a convex cone and that $E \subset \bigcup X_i$. It is easy to finish the proof.

Recall that the *weak uniform structure* on $E$ is associated to the family of all *pseudo-metrics* $d_\ell$, $\ell \in E'$, defined by:

$$(x, y) \mapsto d_\ell(x, y) = |\ell(x - y)|$$

## III.3 Proposition

The weak completion of $E$ is identical to $(E')^*$ (the algebraic dual of $E'$).

**Proof**

$\widehat{E}$, the weak completion of $E$, can be viewed as a subspace of $(E')^*$; notice that $(E')^*$ is weakly complete for the duality with $E'$ and that $\widehat{E}$ is closed in $(E')^*$ for this duality. If $\ell \in E'$ satisfies $\ell = 0$ on $\widehat{E}$ then $\ell = 0$ and the result follows from the Bipolar Theorem proved in I.8.

We are now in a position to prove some elementary but important results.

## III.4 Proposition

For every $\mu \in M^+(E)$, there exists a unique element $r(\mu) \in (E')^*$ such that:

$$\mu(\ell) = \ell(r(\mu)) \text{ for every } \ell \in E'.$$

This element $r(\mu)$ is called the *resultant* of $\mu$.

**Proof**

The restriction of $\mu$ to $E'$ is linear and $r(\mu)$ is nothing but the unique element of $(E')^*$ which determines this linear map.

We introduce now an important set associated to a given positive conical measure.

## III.5 Definition

If $\mu \in M^+(E)$, we set $K_\mu = \{r(\lambda) : \lambda \in M^+(E), 0 \leq \lambda \leq \mu\}$; one has $K_\mu \subset (E')^*$.

## III.6 Proposition

For every $\mu \in M^+(E)$, $K_\mu$ is weakly compact in $(E')^*$ for the duality with $E'$.

**Proof**

The set $\{\lambda : \lambda \in M^+(E), 0 \leq \lambda \leq \mu\}$ is weakly compact for the duality with $h(E)$; since the map $\lambda \mapsto r(\lambda)$ is continuous, when $(E')^*$ is equipped with the duality with $E'$, the result follows.

To study further conical measures it is necessary to introduce the following definition.

## III.7 Definition

Let $X \subset E$ be a cone (not necessarily convex); we say that $\mu \in M^+(E)$ is *supported* by $X$ if, for every $f \in h(E)$, then $\mu(f)$ depends only on the restriction of $f$ to $X$; in other words whenever $f, g \in h(E)$ satisfy $f = g$ on $X$, one has $\mu(f) = \mu(g)$.

This is equivalent to require that ($f \geq 0$ on $X$) implies $\mu(f) \geq 0$. We denote by $M^+(X)$ this set.

Notice that, according to this definition, then $\mu$ is supported by $X$ iff $\mu$ is supported by $\overline{X}$.

## III.8 Proposition

If $\mu \in M^+(E)$ is supported by a weakly complete convex cone $X \subset E$, then $r(\mu) \in X$.

**Proof**

Suppose that $\ell \in E'$ is positive on $X$; then $\mu(\ell) = \ell(r(\mu)) \geq 0$; the result follows from the Bipolar Theorem proved in I.8.

We are now in a position to define the *restriction* of a positive conical measure on $E$ to a cone $X \subset E$.

## III.9 Proposition

Let $X \subset E$ be a cone; if $\mu \in M^+(E)$ there exists a greatest $\lambda \in M^+(E)$, supported by $X$, such that $0 \leq \lambda \leq \mu$; this measure $\mu_X$ satisfies for every $f \in h^+(E)$:

$$\mu_X(f) = \inf\{\mu(g) : g \in h^+(E), g \geq f \text{ on } X\}$$

**Proof**

It suffices to prove that the preceding formula defines a conical measure, namely that $\mu_X$ is additive:

If $f_1, f_2 \in h^+(E)$ and $g_1, g_2 \in h^+(E)$ satisfy $g_1 \geq f_1$ on $X$ and $g_2 \geq f_2$ on $X$ then $g_1 + g_2 \geq f_1 + f_2$ on $X$; it follows that $\mu_X(f_1) + \mu_X(f_2) \geq \mu_X(f_1 + f_2)$.

Conversely, if $\epsilon > 0$ is given, let $g_\epsilon \in h^+(E)$ such that $g_\epsilon \geq f_1 + f_2$ on $X$ and $\mu(g_\epsilon) \leq \mu_X(f_1 + f_2) + \epsilon$; we can assume that $g_\epsilon \leq f_1 + f_2$ and $g_\epsilon = f_1 + f_2$ on $X$. If we set $g_1 = g_\epsilon \wedge f_1$ and $g_2 = g_\epsilon - g_1$ we check that $g_1, g_2 \in h^+(E)$ and that $g_\epsilon = g_1 + g_2$, and that $g_i = f_i$ on $X$ for $i = 1$ and $i = 2$; thus $\mu(g_\epsilon) = \mu(g_1) + \mu(g_2) \geq \mu_X(f_1) + \mu_X(f_2)$ and it follows that $\mu_X$ is indeed a conical measure.

It is easy to see that $\mu_X$ is supported by $X$.

The following approximation lemma is important; recall (Riesz's Decomposition Lemma) that if $x; x_1,,, x_n$ are positive elements contained in a vector lattice with $x \leq \sum_1^n x_i$ there exists $0 \leq y_i \leq x_i, 1 \leq i \leq n$, such that $x = \sum_1^n y_i$.

## III.10 Lemma

If $E$ is weakly complete, then for every $\mu \in M^+(E)$ and every finite sequence $f_1,,, f_n$ in $h(E)$, there exists a finite decomposition of $\mu$ into $\mu = \mu_1 +,,, + \mu_m$, with $\mu_i \geq 0$, such that, if $x_i = r(\mu_i)$, one has, for $1 \leq k \leq n$:

$$\mu(f_k) = \sum_1^m f_k(x_i)$$

**Proof**

By Proposition III.2 there exists a finite family $X_1,,,X_m$ of closed convex cones in $E$, with $E \subset \bigcup X_i$, such that, for every $1 \leq i \leq n$ and every $1 \leq j \leq m$, the function $f_i$ is equal on $X_j$ to some element in $E'$. We check that $\mu \leq \sum_1^m \mu_{X_i}$ (see III.9); since $M(E)$ is a vector lattice (by Riesz's Decomposition Lemma) there exist $\mu_1,,,\mu_m$ in $M^+(E)$ such that

$$\mu = \sum_1^m \mu_i \text{ and } 0 \leq \mu_i \leq \mu_{X_i} \text{ for } 1 \leq i \leq m$$

For every $1 \leq i \leq m$ the conical measure $\mu_i$ is carried by $X_i$, thus $r(\mu_i) \in X_i$; now, it is easy to finish the proof.

## III.11 Proposition

If $E$ is weakly complete, then for every $\mu \in M^+(X)$ the weak closure of the cone generated by $K_\mu$ is the minimal weakly complete convex cone $X$ supporting $\mu$.

**Proof**

If $Y$ is a weakly complete convex cone supporting $\mu$, by Proposition III.8 $K_\mu \subset Y$; conversely, the weak closure of the cone generated by $K_\mu$ is convex and supports $\mu$ by the preceding lemma.

The preceding results were obtained by G. Choquet in the framework of Integral Representation Theory; the following lemma was proved by G. Choquet in order to study the class of *zonoforms*; it will be useful here for a different purpose.

## III.12 Lemma

For every $\mu \in M^+(E)$ and every $\ell \in E'$ one has:

$$\mu(\ell^+) = \sup\{\ell(x) : x \in K_\mu\}$$

**Proof**

For every $0 \leq \lambda \leq \mu$ one has $\lambda(\ell) \leq \lambda(\ell^+) \leq \mu(\ell^+)$, thus $\mu(\ell^+) \geq \sup\{\ell(x) : x \in K_\mu\}$; conversely, by the proof of Lemma III.10, there exists $0 \leq \lambda \leq \mu$, supported by the convex cone $\{\ell \geq 0\}$ such that $\mu(\ell^+) = \lambda(\ell) = \ell(r(\lambda))$, which proves the result.

Now, we prove some results (Lemmas III.14,15) which are useful for our purpose, but which have nothing to do with the Theory of Integral Representation:

## III.13 Definition

An *abstract L space* ($L$-space for short) is a Banach lattice such that, for every $x, y \geq 0$, one has $\|x + y\| = \|x\| + \|y\|$.

It is well-known that every $L$-space is isomorphic (as a Banach lattice) and isometric to a space $L^1(\Omega, \mathcal{A}, m)$.

Let $\mu$ be a conical measure on $E$. If we equip the space $h(E)$ with the semi-norm $N_\mu$ defined by $f \mapsto N_\mu(f) = \mu(|f|)$ we see that, if $|f| \leq |g|$, then $N_\mu(f) \leq N_\mu(g)$; moreover if $f, g \geq 0$ we have $N_\mu(f+g) = N_\mu(f) + N_\mu(g)$; consequently, the Hausdorff completion of $h(E)$, equipped with $N_\mu$, is an $L$-space. There exists a canonical map from $E'$ to $h(E)$; the following key lemma may be viewed as a converse of this property.

## III.14 Lemma

If $u$ is a linear map (not necessarily continuous) from $E'$ to a space $L^1(\Omega, \mathcal{A}, m)$ there exists a lattice homomorphism $p_u$ from $h(E)$ to $L^1(\Omega, \mathcal{A}, m)$ such that, for every $\ell_1, , , \ell_n \in E'$ one has:

$$p_u(\sup(\ell_1, , , \ell_n)) = \sup(u(\ell_1), , , u(\ell_n))$$

[$\sup(\ell_1, , , \ell_n)$ is defined on $E$ by $x \mapsto \sup(\ell_1(x), , , \ell_n(x))$]

Moreover, the map $\mu_u : h(E) \to \mathbb{R}$ defined by $\mu_u(h) = \int p_u(h)dm$ is a positive conical measure on $E$.

**Proof**

We may assume that $E$ is weakly complete, namely $E = (E')^*$; for every $A \in \mathcal{A}$, with $m(A) < \infty$, we define the transposed map $u'(A) \in (E')^*$ by:

$$u'(A)(\ell) = \int_A u(\ell)dm \text{ for every } \ell \in E'$$

Setting $f(x) = \sup(\ell_1(x),,,\ell_n(x))$ for $x \in E$, we prove that:

$$\int_A \sup(u(\ell_1),,,u(\ell_n))dm = \lim \sum f(u'(A_i))$$

where the limit is taken along the filter of all finite partitions of $A$ into $A_1,,,A_m$.

Since $f$ is convex $\sum f(u'(A_i)) \leq \sum f(u'(B_{i,j}))$ whenever, for each $i$ the family $(B_{i,j})$ is a partition of $A_i$.

There exists a partition $A_1,,,A_n$ of $A$ such that

$$\int_A \sup(u(\ell_1),,,u(\ell_n))dm = \sum_1^n \int_{A_i} u(\ell_i)dm = \sum_1^n \ell_i(u'(A_i)) \leq \sum_1^n f(u'(A_i))$$

Conversely, if $B_1,,,B_m$ is any decomposition of $A$ we have

$$\sum_1^m f(u'(B_i)) = \sum_1^m \ell_{k_i}(u'(B_i)) = \sum_1^m \int_{B_i} u(\ell_{k_i})dm$$

where $\ell_{k_i}$ is one of the elements $\ell_1,,,\ell_n$.

Since $\sum_1^m \int_{B_i} u(\ell_{k_i})dm \leq \int_A \sup(u(\ell_1),,,u(\ell_n))dm$, the formula follows.

Now, it is easy to check that $p_u$ induces an homomorphism from $h(E)$ to $L^1(\Omega, \mathcal{A}, m)$.

As it is indicated in the introductory paragraph of this Chapter, taking account of Theorem II.7, we see why conical measures play a key rôle in this book:

With the notations of the above lemma, if $E$ is a Banach space, in order to investigate if $u : E' \to L^1(\Omega, \mathcal{A}, m)$ factors through $L^p(\Omega, \mathcal{A}, m)$ $(1 < p < \infty)$ we need to study the integral

$$\int (\sum_1^n |u(\ell_i)|^p)^{1/p} dm \text{ where } \ell_1,,,\ell_n \in E'$$

If $\mu_u$ is the positive conical measure on $E$ associated to $u$ as in the Lemma III.14, this integral is nothing but

$$\mu_u(h)$$

where $h : E \to \mathbb{R}$ is given by $h(x) = (\sum_1^n |\ell_i(x)|^p)^{1/p}$; generally $h \notin h(E)$ but $\mu_u(h)$ is obviously well defined since $h$ depends only on a finite family of elements of $E'$ and is positively homogeneous.

Since we are interested in order structures the following result will be very useful.

# III.15 Lemma

Let $X \subset E$ be a convex cone and $u : E' \to L^1(\Omega, \mathcal{A}, m)$ be a linear map; the following two properties are equivalent:

1) $\mu_u$ is carried by $X$.

2) $u$ is positive on the polar cone $X^0$ of $X$.

**Proof**

1) $\Rightarrow$ 2): let $\ell \in X^0$; one has $\ell^- = 0$ on $X$, thus $\mu_u(\ell^-) = 0$ and $\int (u(\ell))^- dm = 0$; thus $u(\ell)^- = 0$ and $u$ is positive on $X^0$.

2) $\Rightarrow$ 1): if $\ell \leq 0$ on $X$ one has $u(\ell) \leq 0$, thus $p_u(\ell^+) = 0$ and $\mu_u(\ell^+) = 0$; by Lemma III.12 one has $\ell(x) \leq 0$ for every $x \in K_{\mu_u}$, thus $K_{\mu_u}$ is contained in the closure $\overline{X}$ of $X$ in $(E')^*$; the result follows by Lemma III.11.

Now, we indicate (Proposition III.18) another interpretation of a given conical measure, due to E. Thomas [66], which will be useful in Chapter VI; in some sense this interpretation is dual of the preceding one. To begin with, we show that on every space $\mathcal{C}(K)$, there exists a *canonical* conical measure.

# III.16 Proposition

Let $K$ be a compact Hausdorff space and $\mathcal{C}(K)$ be the space of all (real) continuous functions. We define a conical measure $\Lambda$ on $\mathcal{C}(K)$ by the following formula: for every Radon measures $\mu_1,,,\mu_n$ on $K$, one has:

$$\Lambda(\sup(\mu_1,,,\mu_n)) = \sup(\mu_1,,,\mu_n)(1)$$

[In the left member $\sup(\mu_1,,,\mu_n)$ is the map defined on $\mathcal{C}(K)$ by $f \mapsto \sup(\mu_1(f),,,\mu_n(f))$; in the right member the sup involved is the classical sup of $\mu_1,,,\mu_n$ ]

**Proof**

It is well known that

$$\sup(\mu_1,,,\mu_n)(1) = \sup\{\sum \mu_i(f_i) : f_i \in \mathcal{C}^+(K), \sum f_i = 1\}$$

If we set $h(f) = \sup(\mu_1(f),,,\mu_n(f))$ for $f \in \mathcal{C}(K)$, it is easy to check that $\Lambda(f) = \lim \sum h(g_i)$ along the filter of all positive decompositions $g_1,,,g_m$ in $\mathcal{C}(K)$ with $\sum_1^m g_i = 1$.

Let $E$ and $F$ be two Hausdorff l.c.s. and $\varphi : E \to F$ be a continuous linear map; if $\mu \in M^+(E)$ we define $\varphi(\mu) \in M(E)^+$ by $\varphi(\mu)(h) = \mu(h \circ f)$ for every $h \in h(F)$.

# III.17 Definition

An *abstract M space* ($M$-space for short) is a Banach lattice such that for every $x, y \geq 0$ one has: $\| \sup(x,y)\| \leq \sup(\|x\|, \|y\|)$.

It is well known that, if the unit ball of an $M$-space has a greatest element, then the space is isomorphic (as a Banach lattice) and isometric to a space $\mathcal{C}(K)$, for some Hausdorff compact space $K$.

Let $E$ be a Hausdorff l.c.s. and $\mu \in M^+(E)$; we define $I(\mu) = \{\lambda : \lambda \in M(E), \exists C, |\lambda| \leq C\mu\}$; this is an $M$-space when it is equipped with the norm

$$\|\lambda\| = \inf\{C : |\lambda| \leq C\mu\}$$

Then $\mu$ is the greatest element of the unit ball of $I(\mu)$, thus $I(\mu)$ is isomorphic (as a Banach lattice) and isomorphic to a space $\mathcal{C}(K)$; to $\mu$ is associated the constant function 1. If $K_\mu \subset E$ (in particular if $E$ is weakly complete), there exists a canonical map $\varphi : \mathcal{C}(K) \to E$ given by $\varphi(\lambda) = r(\lambda)$: namely, if $\lambda \in I(\mu)$, denoting again by $\lambda$ the element of $\mathcal{C}(K)$ associated to $\lambda$, we set $\varphi(\lambda)=$ resultant of $\lambda$.

With the preceding notations, we are now in a position to prove the following statement which is, in some sense, dual to III.14.

# III.18 Proposition

Let $\mu \in M^+(E)$ and suppose that $K_\mu \subset E$; then $\mu$ is the image of the (canonical) conical measure $\Lambda$ on $\mathcal{C}(K)$ under the (canonical) map $\varphi : \mathcal{C}(K) \to E$. Briefly:

$$\mu = \varphi(\Lambda)$$

**Proof**

It suffices to apply Lemma III.10.

Now, we deal with a class of convex cones (defined in III.19), which is quite important in the framework of the Theory of Integral Representation, and also in our setting.

## III.19 Notation

We denote by $\mathcal{S}$ the class of all convex subsets (contained in a Hausdorff l.c.s.) which are weakly complete and contain no line; for a convex cone $X$, the latter condition is equivalent to $X \cap -X = \{0\}$ ($X$ is said to be *proper*).

The following lemma will enable us to establish a connection with the class of normal cones contained in a Banach space (Proposition V.2).

## III.20 Lemma

If $X \subset E$ is a cone in $\mathcal{S}$, then every $\ell \in E'$ admits a decomposition $\ell = \ell_1 - \ell_2$, where $\ell_1, \ell_2 \in E'$ are positive on $X$.

In other words $X$ can be embedded in a cone $(\mathbb{R}^+)^I$ for some set $I$.

**Proof**

Let $X^0$ be the polar cone of $X$ in $E'$, namely $X^0 = \{\ell : \ell \in E', \ell \geq 0 \text{ on } X\}$; we claim that $E' = X^0 - X^0$, otherwise there would exist $\varphi \in (E')^*$ with $\varphi \neq 0$ and $\varphi(\ell) = 0$ for every $\ell \in X^0$, consequently $\varphi$ belongs to the bipolar of $X$ taken in $(E')^*$, for the duality with $E'$. Since $X \in \mathcal{S}$ it follows that $X$ is closed in $(E')^*$ for the duality with $E'$, thus $\varphi \in X$; in the same way $\varphi \in -X$; since $X$ is assumed to be proper one has $\varphi = 0$ and the first part of the result follows.

To prove the second assertion it suffices to choose for $I$ the set of all $\ell \in E'$ which are positive on $X$.

The link between the class $\mathcal{S}$ and the class of normal cones (contained in a Banach space) will be made precise in Proposition V.2; however this link can be guessed if one compares Lemma III.20 and Theorem I.13.

The preceding result has the following three consequences.

## III.21 Proposition

If $X \subset E$ is a cone in $\mathcal{S}$ then for the weak (relative) topology on $X$ the origin has a basis of neighbourhoods consisting of sets of the form $\{x : x \in X : \ell(x) < 1\}$ where $\ell \in E'$ satisfies $\ell \geq 0$ on $X$.

**Proof**

By definition the family of sets $\{x : x \in X,\ |\ell_1(x)| < 1,,, |\ell_n(x)| < 1\}$, for all finite sequences $\ell_1,,, \ell_n$ in $E'$, is a basis of neighbourhoods of 0 in $X$; by the preceding lemma every $\ell_i$ can be written $\ell_i = \alpha_i - \beta_i$ where $\alpha_i, \beta_i \in E'$ are positive on $X$, it suffices to take $\ell = 2\sum(\alpha_i + \beta_i)$ to get a neighbourhood of 0 in $X$ contained in the given one.

## III.22 Proposition

If $X \subset E$ is a cone in $\mathcal{S}$, for every $f \in h(E)$ there exists $\ell \in E'$ such that $f(x) \leq \ell(x)$ for every $x \in X$.

**Proof**

It suffices to apply Lemma III.20.

## III.23 Proposition

Let $X \subset E$ be a cone in $\mathcal{S}$; for $f \in h(E)$ set:

$$\widehat{f}(x) = \inf\{\ell(x) : \ell \in E', \ell \geq f \text{ on } X\} \text{ for every } x \in E$$

Then $\widehat{f}$ is real valued; it is the smallest upper semicontinuous (u.s.c.) superlinear function on $X$, greatear than $f$, and one has:

$$\widehat{f}(x) = \sup\{\mu(x) : \mu \in M^+(X), r(\mu) = x\} \text{ for every } x \in X$$

**Proof**

By Lemma III.20 $\widehat{f}$ is real valued; let $g(x)$ be the right member of the equality we are aiming for; obviously $g$ is superlinear and $g \geq f$ on $X$; it suffices to prove that $g$ is u.s.c.: let $x_0 \in X$ and let $\mathcal{U}$ be an ultrafilter on $X$ converging to $x_0$ and $\epsilon > 0$; for every $x \in X$ let $\mu_x \in M^+(X)$ with $r(\mu_x) = x$ such that $\mu_x(f) \geq g(x) - \epsilon$; it suffices to check that $\mu_x$ converges along $\mathcal{U}$ (for the duality with $h(E)$) towards some $\mu \in M^+(X)$ with $r(\mu) = x_0$ and $\mu(f) \geq \lim_{\mathcal{U}} g(x) - \epsilon$. Since every $h \in h(E)$ is majorized by some $\ell \in E'$ which is positive on $X$ one has $\mu_x(h) \leq \ell(x)$ and since $x \to x_0$ along $\mathcal{U}$ it follows that $\mu_x(h)$ converges along $\mathcal{U}$ to the required $\mu(h)$, consequently $g(x_0) \geq \lim(g(x))$ and $g$ is u.s.c..

## III.24 Notes and Remarks.

1) As said in the introductory paragraph of this Chapter, conical measures were introduced in 1962 by G. Choquet in the framework of the Theory of Integral Representation; the main role is then played by the cones in

$\mathcal{S}$ and the Choquet Order, denoted by $\prec$, on $M^+(X)$, where $X \subset E$ is a given cone in $\mathcal{S}$, namely:

$$(\lambda \prec \mu) \text{ iff } \lambda(f) \leq \mu(f) \text{ for every } f : x \mapsto \sup(\ell_1(x),,,\ell_n(x))$$

where $\ell_1,,,\ell_n \in E'$

In the Posface of our book [14] G. Choquet related the thought process which led him to find out the class $\mathcal{S}$ and conical measures; in the present book we do not use this order (except in Appendix 3).

Some years after, in 1968, G. Choquet (see [14], $V$) took advantage of conical measures to study the class of *zonoforms*: by definition a zonoform is the image of a set $K_\mu$ under a translation.

Conical measures are also involved in two other theories:

In 1974 L. Le Cam [38], [39] pointed out that conical measures fit in the formalism of Statistical Decision Theory; in particular the order $\prec$, on the set of positive conical measures supported by a cone $(\mathbb{R}^+)^I$, plays an important role.

In 1976 I. Kluvanek [32] used conical measures to study the ranges of *vector measures* in the framework of Control Systems; in particular, he associated a conical measure to every vector measure.

However, notice that conical measures are not involved in the Handbook of Measure Theory.

2) In his book B. Maurey [44] used only *cylindrical measures*: if $E$ is a Hausdorff l.c.s., let $a(E)$ be the vector lattice of functions on $E$ generated by $E'$ and the constant functions and $b(E)$ be the subset of $a(E)$ consisting of all bounded functions. By definition, a cylindrical measure $\gamma$ on $E$ is a positive linear form on $b(E)$ such that $\gamma(f_n) \to 0$ for every sequence $(f_n)$ in $b(E)$ such that:

a) $f_n$ decreases to 0 on $E$.

b) There exist $\ell_1,,,\ell_k$ in $E'$ such that, for every $n$, $f_n$ belongs to the vector lattice (of functions on $E$) generated by $\ell_1,,,\ell_k$ and the constant functions.

This presentation of cylindrical measures in due to G. Choquet ([14], VI.1); when $\gamma(1) = 1$ we say that $\gamma$ is a *cylindrical probability.*

A result, similar to Lemma III.14, asserts that:

Let $L^0(\Omega, \mathcal{A}, P)$ be the space of all (almost everywhere finite) measurable real functions on a measurable space $(\Omega, \mathcal{A}, P)$, where $P$ is a probability.

If $E$ is a Hausdorff l.c.s. and if $u$ is a (not necessarily continuous) linear map from $E'$ to $L^0(\Omega, \mathcal{A}, P)$ there exists a lattice homomorhism $p_u$ : $a(E) \to L^0(\Omega, \mathcal{A}, P)$ which satisfies:

$$p_u(1) = 1 \text{ and } p_u(\ell) = u(\ell) \text{ for every } \ell \in E'$$

Consequently, the map $\gamma_u(f) = \int p_u(f)dP$, for $f \in b(E)$, is a cylindrical probability on $E$.

Let $E$ be a Banach space, and suppose $u(E) \subset L^q(\Omega, \mathcal{A}, P)$, where $1 < q < \infty$. If we want to know whether $u$ factors through $L^p(\Omega, \mathcal{A}, P)$, for some $q < p < \infty$, in the sense of II.10.4), we need to study the integral:

$$\int (\sum_1^n |u(\ell_i)|^p)^{q/p} dP \text{ where } \ell_1,,,\ell_n \in E'$$

This integral is nothing but $\gamma_u(f)$ where $f(x) = (\sum_1^n |\ell_i(x)|^p)^{q/p}$ for $x \in E$.

Since $f$ is not positively homogeneous of degree 1, when $q > 1$, we cannot work with a conical measure, and we need to use cylindrical measures. On the other hand, when $q = 1$, the function $f$ is positively homogeneous of degree 1 and the *cylindrical algebra* of $E$ (the algebra of subsets of $E$ generated by the elements of $E'$) is too big, because it does not enable us to deal with convex subcones of $E$. In fact, as we shall see in Chapter IV, conical measures are exactly the tool which is needed to cope with convex cones contained in a Banach space.

# Chapter IV

# Structure of convex cones contained in a Banach space

To every convex cone $X$, contained in a Banach space $B$, we associate (in IV.1) an *index* $i(X)$ which will be useful all along this book; we emphasize that the convex cone $X$ is not assumed to be proper. This index is involved in two purposes:

1) To study operators defined on $B$ which are *p-summing on* $X$, in the sense of Definition 6 of the Introduction. Note that Chapter VII is devoted to a systematic study of this class of operators, but, in this Chapter, we shall use only very basic results.

2) To study the factorization (in the sense of Definition II.1) of operators defined on $B'$, with values in a space $L^1(\Omega, \mathcal{A}, m)$, which are assumed to be positive on the polar cone $X^0 \subset B'$.

When $X$ is the whole Banach space $B$ (of infinite dimension) we are in the framework of the classical Maurey Theory [44], and $i(X)$ is nothing but the cotype index $q_B$ of $B$ (see Theorem IV.2).

The main results are Theorems IV.10 and IV.15; to prove Theorem IV.15 we use Rosenthal's Lemma proved in II.8.

$B$ will always stand for a Banach space of infinite dimension and $B_1$ for its closed unit ball.

Let us repeat the definition of the index $i(X)$, given previously in the Introduction (Def. 7); $M_p$ (of a sequence) is defined in the Introduction (Def. 5): namely if $(x_i)$ is a sequence in $B$ and $1 \leq p \leq \infty$, then:

$$M_p((x_i)) = \sup\{(\sum_i |x'(x_i)|^p)^{1/p} : x' \in B', \|x'\| \leq 1\}$$

# IV.1 Definition

If $X \subset B$ is a convex cone (not necessarily proper) we set:

$$i(X) = \inf\{q : q \geq 1,\ \exists C,\ (\sum_1^n \|x_i\|^q)^{1/q} \leq CM_1((x_i))\}$$

for every finite sequence $x_1,,,x_n$ in $X$.

When $X$ is the whole Banach space $B$ the following result is due to B. Maurey and G. Pisier ([46] Prop. I.4); the definition of the *cotype index* $q_B$ of $B$ is recalled in Appendix 1 where we have gathered several results concerning Banach spaces.

# IV.2 Theorem

For every Banach space $B$ of infinite dimension one has $i(B) = q_B$.

Before going further, we need some results concerning conical measures on Banach spaces; we refer to III.14 for the definitions.

# IV.3 Proposition

Let $B$ be a Banach space and $B'$ its dual. For a linear map $u$ from $B'$ to a space $L^1(\Omega, \mathcal{A}, m)$ the following two properties are equivalent:

1) $u$ is continuous.

2) $K_\mu \subset B''$, where we let stand $\mu$ for $\mu_u$.

In this case one has: (diameter of $K_\mu$) $\leq \|u\| \leq 2$(diameter of $K_\mu$).

**Proof**

1) $\Rightarrow$ 2): if $\lambda \in M^+(B)$ satisfies $0 \leq \lambda \leq \mu_u$, for every $\ell \in B'$ one has

$$|\lambda(\ell)| \leq \lambda(|\ell|) \leq \mu_u(|\ell|) = \int |u(\ell)|dm \leq \|u\|\|\ell\|$$

Thus, the map $\ell \mapsto \ell(r(\lambda))$ is continuous, $r(\lambda) \in B''$ and $K_\mu \subset B''$. Moreover, from Lemma III.12 we get (diameter of $K_\mu$) $\leq \|u\|$; the inequality $\|u\| \leq$ 2(diameter of $K_\mu$) is proved in the same way.

2) $\Rightarrow$ 1): since $K_\mu$ is weakly compact in $B''$ for the duality with $B'$, $K_\mu$ is bounded in $B''$; the result follows again from Lemma III.12.

Taking into account the preceding proposition, it is natural to introduce the following.

## IV.4 Definition

A conical measure $\mu \in M^+(B)$ is said to be *continuous* if $K_\mu \subset B''$; in this case we set

$$\|\mu\| = \sup\{\mu(|\ell|) : \ell \in B'_1\}$$

We are now in a position to rephrase Theorem II.7 in the setting of conical measures.

## IV.5 Proposition

If $u : B' \to L^1(\Omega, \mathcal{A}, m)$ is a continuous linear map, for every $1 < p < \infty$ the following two properties are equivalent:

1) $u$ factors through $L^p(\Omega, \mathcal{A}, m)$.

2) For every finite sequence $\ell_1, , , \ell_n$ in $B'$, denoting by $f$ the function defined on $B$ by $x \mapsto (\sum_1^n |\ell_i(x)|^p)^{1/p}$, one has:

$$\mu_u(f) \leq C(\sum_1^n \|\ell_i\|^p)^{1/p}$$

where $C$ is some constant.

Moreover, when these conditions are satisfied, the smallest possible constant $C$ is equal to $C_{1,p}(u)$.

**Proof**

$\mu_u(f)$ is an abuse of notation, since generally $f \notin h(B)$, but it is easy to see that for every $\lambda \in M^+(B)$, the integral $\lambda(f)$ is well defined; one checks that $\mu_u(f) = \int (\sum_1^n |u(\ell_i)|^p)^{1/p} dm$.

## IV.6 Notation

Let $\mu \in M^+(B)$; by the preceding result, III.13 and III.14 the notation $C_{1,p}(\mu)$ makes sense whenever the canonical map from $h(E)$ to $L^1(\mu)$, which can be viewed as a space $L^1(\Omega, \mathcal{A}, m)$, factors through $L^p(\Omega, \mathcal{A}, m)$.

In this case, by abuse of notation, we say that $\mu$ *factors* through $L^p$.

A classical theorem of B. Maurey and G. Pisier (see Theorem IV.2) relating the theory of cotype to the theory of factorization, can be rephrased in the setting of conical measures:

## IV.7 Theorem

Let $q'_B$ satisfy $1/q'_B + 1/q_B = 1$; if $q_B$ is finite, then for every $1 < p < q'_B$ there exists $C_p$ such that, for every continuous conical measure $\mu \in M^+(B)$, for every finite sequence $\ell_1,,,\ell_n$ in $B'$, denoting by $f$ the map defined on $B$ by $x \mapsto (\sum_1^n |\ell_i(x)|^p)^{1/p}$, one has:

$$\mu(f) \leq C_p \|\mu\| (\sum_1^n \|\ell_i\|^p)^{1/p}$$

Conversely, every such $p$ satisfies $1 \leq p \leq q'_B$.

The following lemma of approximation will be useful in the sequel.

## IV.8 Lemma

Let $X \subset B$ be a convex cone and $\widehat{X}$ be its closure in $B''$ for $\sigma(B'', B')$. If $x_1,,,x_n$ is a finite sequence in $\widehat{X}$, there exists a directed set of sequences $x_1^\alpha,,,x_n^\alpha$ in $X$, $\alpha \in \mathcal{U}$, such that, for every $\alpha \in \mathcal{U}$, $M_1((x_i^\alpha)) \leq M_1((x_i))$ and $x_i^\alpha \to x_i$ along $\mathcal{U}$ for $\sigma(B'', B')$ for $1 \leq i \leq n$.

**Proof**

For every $\epsilon \in \mathcal{E} = \{-1, 1\}^n$ let $K_\epsilon = \{(\epsilon_1 \ell,,,\epsilon_n \ell) : \ell \in B'_1\}$; $K_\epsilon$ is convex and compact in $(B')^n$ for the duality with $B^n$ and the same holds for the convex hull of $\cup K_\epsilon$ (where $\epsilon \in \mathcal{E}$) denoted by $K$.

Consequently, the set $E = (X^0)^n + K$ is closed in $(B')^n$ for the duality with $B^n$ since it is the sum of a closed set and of a compact set; moreover, if $\overline{X}$ is the closure of $X$ in $B$, the polar $E^0$ of $E$ in $B^n$ is

$$E^0 = \{(y_1,,,y_n) : (y_1,,,y_n) \in (\overline{X})^n,\ M_1((y_i)) \leq 1\}$$

Since $E$ is closed in $(B')^n$ for the duality with $B^n$ it follows that $E$ is the polar of $E^0$; the proof is completed by checking that the polar of $E$ in $(B'')^n$ is

$$\{(z_1,,,z_n) : (z_1,,,z_n) \in (\widehat{X})^n,\ M_1((z_i)) \leq 1\}$$

we use here the fact that $M_1((z_i)) = \sup\{\sum_1^n |\ell(z_i)| : \ell \in B'_1\}$.

The following lemma provides a useful family of $p$-summing operators from $B$ to $\ell^p(\mathbb{N})$; the definition of $p$-summing operators on a convex cone is given in the Introduction (Definition 6) and Chapter VII is devoted to this class of operators: namely, if $X \subset B$ is a convex cone (not necessarily proper) and if $p \geq 1$, a linear operator $T$ (not necessarily continuous), from $B$ to another

Banach space $F$, is said to be *p-summing on* $X$ if for some $C > 0$ one has for every $x_1,,,x_n \in X$:

$$(\sum_1^n \|T(x_i)\|^p)^{1/p} \leq CM_p((x_i))$$

We denote by $\pi_p(X,T)$ the minimal constant $C$ and by $\Pi_p(X;B,F)$ the set of these operators.

## IV.9 Lemma

Let $1 \leq p < \infty$ and $(\ell_n)$ be a sequence in $B'$ with $(\sum_n \|\ell_n\|^p)^{1/p} < \infty$; the map $T: B \to \ell^p(\mathbb{N})$ defined by $x \mapsto (\ell_n(x))$ satisfies:

$$\|T\| = M_p((\ell_n)) \text{ and } \pi_p(B,T) \leq (\sum_n \|\ell_n\|^p)^{1/p}$$

**Proof**

Concerning $\|T\|$ it suffices to check that $M_p((\ell_n)) = \sup\{(\sum_n |\ell_n(x)|^p)^{1/p} : x \in B_1\}$.

If $x_1,,,x_n$ is a finite sequence in $B$ one has

$$\sum_i \|T(x_i)\|^p = \sum_{i,j} |\ell_j(x_i)|^p \leq (\sum_j \|\ell_j\|^p) M_p((x_i))^p$$

which is the result we are aiming for.

We are now in a position to prove the first main result of this Chapter; we abbreviate $\ell^p(\mathbb{N})$ to $\ell^p$.

## IV.10 Theorem

Let $X \subset B$ be a convex cone (not necessarily proper); for $1 < p < \infty$ the following four properties are equivalent:

a) There exists a constant $C_a$ such that, for every scalarly $\ell^1$-sequence $(x_n)$ in $X$, there exist a scalarly $\ell^p$-sequence $(y_n)$ in $X$ and a norm 1 positive sequence $(\alpha_n)$ in $\ell^{p'}$ such that $x_n = \alpha_n y_n$ for every $n$ and

$$M_p((y_n)) \leq C_a M_1((x_n))$$

b) There exists a constant $C_b$ such that, for every Banach space $F$, if $v \in \Pi_p(X; B, F)$ one has also $v \in \Pi_1(X; B, F)$ and

$$\pi_1(X; B, F) \leq C_b \pi_p(X; B, F)$$

c) There exists a constant $C_c$ such that for every continuous conical measure $\mu \in M^+(B, B')$, supported by $X$, for every finite sequence $\ell_1, , , \ell_n$ in $B'$, denoting by $f$ the function defined on $B$ by $x \mapsto (\sum_1^n |\ell_i(x)|^p)^{1/p}$, one has

$$\mu(f) \leq C_c \|\mu\| (\sum_1^n \|\ell_i\|^p)^{1/p}$$

d) There exists a constant $C_d$ such that, for every space $L^1(\Omega, \mathcal{A}, m)$, every operator $u : B' \to L^1(\Omega, \mathcal{A}, m)$, positive on the polar cone $X^0 \subset B'$, factors through $L^p(\Omega, \mathcal{A}, m)$ and satisfies

$$C_{1,p}(u) \leq C_d \|u\|$$

When the preceding conditions are satisfied the identity map is $(p', 1)$-summing on $X$, namely there exists a constant $C_p$ such that, for every $x_1, , , x_n$ in $X$ one has:

$$(\sum_1^n \|x_i\|^{p'})^{1/p'} \leq C_p M_1((x_i))$$

**Proof**

a) $\Rightarrow$ b) with $C_b \leq C_a$: for every finite sequence $x_1, , , x_n$ in $X$, one has

$$\sum_1^n \|v(x_i)\| = \sum_1^n \alpha_i \|v(y_i)\| \leq (\sum_1^n \|v(y_i)\|^p)^{1/p}$$

by Holder's Inequality; the result follows since:

$$(\sum_1^n \|v(y_i)\|^p)^{1/p} \leq \pi_p(X, v) M_p((y_i)) \leq \pi_p(X, v) C_a M_1((x_i))$$

b) $\Rightarrow$ c) with $C_c \leq C_b$: there are three steps.

$\alpha$) The function $f$ can be extended in an obvious way to $B''$; by abuse of notation we continue to write $f$. Then $f$ is the pointwise supremum on $B''$ of the set of all the functions $\ell \in B'$ which are majorized by $f$. For every $\epsilon > 0$ there exists $0 \leq f_\epsilon \leq f$, which is a finite supremum of elements of $B'$, such that $\mu(f_\epsilon) \geq \mu(f) - \epsilon$.

$\beta$) By Lemma III.10 there exists a decomposition $\mu = \sum_1^k \mu_i$ with $\mu_i \geq 0$, such that, if $x_i = r(\mu_i)$, one has $\mu(f_\epsilon) = \sum_1^k f_\epsilon(x_i)$; moreover $x_i \in B''$ because $\mu$ is continuous.

We have

$$M_1((x_i)) = \sup\{\sum_1^k |\ell(x_i)| : \ell \in B_1'\} \leq \sup\{\sum_1^k \mu_i(|\ell|) : \ell \in B_1'\}$$

thus $M_1((x_i)) \leq \|\mu\|$.

By Lemma IV.8, since $\mu$ is supported by $X$, there exist $z_1,,,z_n \in X$ such that $M_1((x_i)) \leq \|\mu\|$ and

$$|\sum_1^k f_\epsilon(z_i) - \sum_1^k f_\epsilon(x_i)| \leq \epsilon$$

$\gamma$) We now apply property (b) to the map $v : B \to \ell^p$ defined by

$$v(x) = (\ell_1(x),,,\ell_n(x),0,,,)$$

and to the sequence $z_1,,,z_k$ in $X$.

For every $x \in B$ one has $\|v(x)\| = f(x)$ and, by the preceding lemma one has:

$$\pi_p(X,v) \leq \pi_p(B,v) \leq (\sum_1^n \|\ell_i\|^p)^{1/p}$$

We get

$$\sum_1^k f(z_i) \leq C_b(\sum_1^n \|\ell_i\|^p)^{1/p} M_1((z_i))$$

The result follows, with $C_c \leq C_b$, using $\alpha$) and $\beta$).

c) and d) are equivalent with $C_c = C_d$ by Lemma III.15 and Proposition IV.5.

d) $\Rightarrow$ a) with $C_a \leq C_d$: if $(x_i)$ is a scalarly $\ell^1$-sequence in $B$, the map $u : B' \to \ell^1$ defined by $u(\ell) = (\ell(x_i))$ satisfies $\|u\| = M_1((x_i))$; to conclude the proof it suffices to apply property d).

To prove that the identity map is $(p',1)$-summing on $X$ we use property a); with the notations of a) one has:

$$(\sum_1^n \|x_i\|^{p'})^{1/p'} \leq \sup\{\|y_n\| : n \in \mathbb{N}\} \leq \sup\{\|x_n\| : n \in \mathbb{N}\} \leq C_a M_1((x_i))$$

## IV.11 Remark

In the preceding theorem, condition a) depends only on $X \cap B_1$, because, since

$$M_p((x_i)) = \sup\{\|\sum a_i x_i\| : (\sum |a_i|^{p'})^{1/p'} \leq 1\}$$

we get

$$(1/2)M_p((x_i)) \leq \sup\{\|\sum a_i x_i\| : a_i \geq 0, (\sum a_i^p)^{1/p} \leq 1\} \leq M_p((x_i))$$

In other words, if $\tilde{X}$ is another convex cone contained in a Banach space $\tilde{B}$ and if there exists an affine one-to-one map from $X$ onto $\tilde{X}$ such that $X \cap B_1$ and $\tilde{X} \cap \tilde{B}_1$ are absorbing each other, then a) holds for $X$ iff it holds for $\tilde{X}$.

It is now natural to introduce the following

## IV.12 Notation

We denote by $I_X$ the set of all $p > 1$ satisfying the conditions of Theorem IV.10.

It will be shown in Proposition VII.5 (Pietsch's Inequality) that, if an operator on a Banach space $B$ is $p$-summing ($1 \leq p < \infty$) on some convex cone $X \subset B$ then it is also $q$-summing on $X$ for every $q \geq p$; consequently we have the following

## IV.13 Proposition

$I_X$ is an interval of $\mathbb{R}$ with left extremity 1.

In the preceding proposition we do not know whether the right extremity belongs to $I_X$ or not.

## IV.14 Notation

We denote by $I(X)$ the right extremity of $X$; if $I_X = \emptyset$, we set $I(X) = 1$.

By Theorem IV.10, if $p \in I_X$ then the identity is $(p', 1)$-summing on $X$, in other words $p' \geq i(X)$; now, we prove a converse of this property.

# IV.15 Theorem

One has $I(X) = \sup\{p : \text{the identity is } (p',1)\text{-summing on } X\}$, in other words

$$1/I(X) + 1/i(X) = 1$$

More precisely, if $p \notin I_X$ the identity is not $(p',1)$-summing on $X$.

This result will be deduced from the following lemma; its proof involves Rosenthal's Lemma proved in III.8.

# IV.16 Lemma

Let $p > I(X)$; for every integer $n$ and $\epsilon > 0$ there exist norm 1 elements $x_1,,,x_n \in X$ such that for all scalars $a_1,,,a_n$ one has

$$\|\sum_1^n a_i x_i\| \leq (2+\epsilon)(\sum_1^n a_i^{p'})^{1/p'}$$

**Proof**

By a) of Theorem IV.10, for every $C$ there exists a scalarly $\ell^1$-sequence $(x_i)$ in $X$ such that, if we denote by $\mu$ the conical measure $\sum_i \epsilon_{x_i}$, one has $C_{1,p}(\mu) > C\|\mu\|$, where, by abuse of notation we set $C_{1,p}(\mu) = +\infty$ if $\mu$ fails to be $p$-factorizable. Thus, by Theorem IV.5, there exists a finite sequence $\ell_1,,,\ell_m$ in $B'$ such that, denoting by $f$ the map defined on $B$ by $x \mapsto (\sum_1^m |\ell_i(x)|^p)^{1/p}$, one has

$$\mu(f) > C\|\mu\|(\sum_1^m \|\ell_i\|^p)^{1/p}$$

There exists an integer $k$ such that if $\mu_k = \sum_1^k \epsilon_{x_i}$ one has

$$\mu_k(f) > C\|\mu\|(\sum_1^m \|\ell_i\|^p)^{1/p} \text{ , thus } \mu_k(f) > C\|\mu_k\|(\sum_1^m \|\ell_i\|^p)^{1/p}$$

in other words we can assume that the sequence $(x_i)$ is eventually 0.

Now, we shall apply Rosenthal's Lemma (II.8); to fit with the notations of this lemma we set

$$\Omega = \{x_1,,,x_k\} \text{ and } m = \sum_1^k \epsilon_{x_i}$$

It is clear that we can assume that the $x_i$ are all distinct and non-zero for $1 \leq i \leq k$.

Define $u : B' \to L^1(\Omega, m)$ by $u(\ell) = (\ell(x_i))$; we can assume that $\|u\| = 1$, in other words that $M_1((x_1,,,x_k)) = 1$.

One has $C_{1,p} > C$ by the way things were set up.

By Rosenthal's Lemma, if $n$ and $0 < \delta < 1$ are given, if $C$ is big enough (depending only on $n$ and $\delta$) there exists a probability $P = \sum_1^k \alpha_i \epsilon_{x_i}$ on $\Omega$, with $\alpha_i > 0$ for $1 \leq i \leq k$ because $x_i \neq 0$ for every $i$, such that the map $u_1 : B' \to L^1(\Omega, P)$ defined by $u_1(\ell) = (\ell(x_i)/\alpha_i)$ satisfies:

a) $\|u_1\| = 1$.

b) $\int |u_1(\ell)|^p dP \leq C^p \|\ell\|^p$ for every $\ell \in B'$.

c) There exist pairwise disjoint sets $A_i \subset \Omega$, $1 \leq i \leq n$, and norm 1 elements $\ell_i \in B'$, $1 \leq i \leq n$, such that for $1 \leq i \leq n$ one has:

$$\int_{A_i} |u_1(\ell_i)|^p dP \geq \delta^p C^p$$

The rest of the proof runs as follows:

For every $1 \leq i \leq n$, by c) there exists $\varphi_i \in L^{p'}(\Omega, m)$, $\varphi_i \geq 0$ of norm 1, supported by $A_i$ and such that:

d) $\int_{A_i} u_1(\ell_i)\varphi_i dP \geq \delta C/2$ (may be it is necessary to change $\ell_i$ for $-\ell_i$).

Define $v : B' \to \ell_n^p$ by $v(\ell) = (1/C) \int u_1(\ell)\varphi_i dP$, in the canonical basis $(e_1,,,e_n)$ of $\ell_n^p$; if $v(\ell)_i$ is the $i$-coordinate of $v(\ell)$, one has

$$|v(\ell)_i|^p \leq (1/C^p) \int_{A_i} |u_1(\ell)|^p dP$$

By b) and since the sets $A_i$ are pairwise disjoint one has $\|v\| \leq 1$; it is easy to check that for $1 \leq i \leq n$ the element $v^*(e_i^*)$ is of the form $v^*(e_i^*)(\ell) = \ell(z_i)$ for every $\ell \in B'$ and for some $z_i$ contained in the convex cone generated by $x_1,,,x_n$, because $\varphi_i$ is positive; in particular $z_i \in X$. Since $\|u\| \leq 1$ for all $a_1,,,a_n \in \mathbb{R}$ one has:

$$\|\sum_1^n a_i z_i\| \leq (\sum_1^n |a_i|^{p'})^{1/p'}$$

By d) one has $\ell_i(z_i) \geq \delta/2$, thus $\|z_i\| \geq \delta/2$, which prove the lemma.

**Proof of Theorem IV.15:**

With the notations of Lemma IV.16 one has $M_1((x_i)) \leq 3n^{1/p'}$, consequently, for every $p_1 > p$, the identity is not $(p'_1, 1)$-summing on $X$, because a relation $n^{1/p'_1} \leq Cn^{1/p'}$ cannot hold when $n$ is big enough.

Let us indicate some properties of the index $i(X)$.

## IV.17 Proposition

One has $i(X) \leq q_B$, where $q_B$ is the cotype index of $B$.

**Proof**

It suffices to use definition IV.15 and the following result of B. Maurey and G. Pisier, recalled in IV.2,

$$q_B = \inf\{q : \text{ the identity is } (q,1) - \text{summing on } B\}$$

From Definition IV.1, it follows that:

## IV.18 Proposition

If $X$ and $Y$ are two convex cones contained in $B$ with $X \subset Y$ one has $i(X) \leq i(Y)$.

If $X$ and $Y$ are two convex cones contained in $B$ one has $i(X+Y) \geq \sup\{i(X), i(Y)\}$.

Let us discuss some examples

## IV.19 Example

We have mentioned in the introductory paragraph of this Chapter that, by a theorem of B. Maurey and G. Pisier, for every Banach space of infinite dimension $i(B)$ is nothing but the cotype index $q_B$ of $B$. It is well known that $q_B \geq 2$.

## IV.20 Example

When the convex cone $X$ is in some sense very thin one has $i(X) = 1$; in particular this is the case if there exist an affine function $\varphi$ on $X$ and a constant $C$ such that

$$\text{for every } x \in X \text{ one has } \|x\| \leq \varphi(x) \leq C\|x\|$$

## IV.21 Example

If $B$ is an infinite dimensional $L^p$ space, $1 \leq p \leq \infty$, and $X = L^p_+$ then $i(X) = p$:

If $f_1,,,f_n$ is a finite sequence in $X$ one has $M_1((f_i)) = \|\sum_1^n f_i\|$ and:

$$\sum_1^n \|f_i\|^p \leq \|\sum_1^n f_i\|^p \text{ because } \sum_1^n \int f_i^p \leq \int (\sum_1^n f_i)^p$$

thus $i(X) \leq p$.

If, moreover, $f_1,,,f_n$ are norm one pairwise disjoint one has

$$M_1((f_i)) = n^{1/p}$$

It follows that $i(X) = p$.

## IV.22 Remark

It is well known that for an $L^p$ space $B$ (of infinite dimension) one has:

$q_B = p$ if $p \geq 2$.

$q_B = 2$ if $p \leq 2$.

With the notations of the preceding example when $1 \leq p < 2$ one has

$$B = X + (-X) \text{ but } i(B) = 2 \text{ and } i(X) = i(-X) = p$$

In this case one has $i(X + (-X)) > \sup\{i(X), i(-X)\}$.

## IV.23 Example

Let $B$ be the Sobolev space $W_0^{1,p}$ consisting of all absolutely continuous functions $f$ on $I = [0, 1]$, such that $f(0) = f(1) = 0$ and $f' \in L^p$, equipped with the norm $\|f\| = \|f'\|_p = (\int |f'(x)|^p dx)^{1/p}$ and let $X \subset B$ be the convex subcone consisting of all positive functions. It is clear (using derivation) that $B$ is isometric to the subspace of $L^p(I, dx)$ consisting of all $u$ such that $\int_0^1 u(t)dt = 0$. Consequently, if $p \geq 2$, since the cotype index of $L^p$ is $p$, one has $i(X) \leq p$. To prove the reverse inequality let $f_1, , , f_n$ be a finite sequence in $X$ with $f_i \wedge f_j = 0$ whenever $i \neq j$; one has

$$M_1((f_i)) = M_1((f_i')) \text{ taken in } L^p(I, dx)$$

thus $M_1((f_i)) = \| \sum_1^n f_i \| = (\sum_1^n \|f_i'\|_p^p)^{1/p}$
If we choose every $f_i$ of norm one we get that $i(X) = p$.

## IV.24 Notes and remarks

The results of this Chapter were obtained in [7], [9] in the case when the cone $X$ is assumed to be normal and extended in [13] to the general case. The main reason why we had first limited ourself to the framework of normal cones is the following: as we shall see in the next Chapter there is a strong connection between the class of normal cones and the one of weakly complete convex cones and in such a cone every positive conical measure has a resultant (namely the resultant is contained in the cone). In fact, as we have seen in this Chapter, using Lemma IV.18 we can deal with an arbitrary convex cone.

# Chapter V

# The case of normal cones

In this Chapter we continue the investigation of convex cones, in the setting of normal cones. We study the relationships between the class of normal cones and the one of cones in $\mathcal{S}$ (Proposition V.2) and we precize Theorem IV.10 in the case when the cone involved is normal (Theorem V.6). Then we prove that if $X$ is a normal cone with $i(X) > 1$, contained in a Banach space $B$, then $i(X)$ is the biggest $r$ such that $B$ contains uniform copies of spaces $\ell_n^r$ whose unit vectors are contained in $X$ (Theorems V.15,17). It follows that, if $i(X) > 1$ then $p_B \leq i(X) \leq q_B$ where $p_B$ and $q_B$ are the type index and cotype index of $B$. We study the particular case when the cone $X \subset B$ is closed and both normal and generating (Theorem V.25). Finally, we study some classes of operators related to a normal cone $X \subset B$ (Theorem V.29).

We use Krivine's Theorem, alluded to at the end of section C of our Introduction, to prove Theorem V.15.

In this Chapter, $B$ stands for a Banach space of infinite dimension.

For convenience, we repeat here the definition of normal cones.

## V.1 Definition

A convex cone $X \subset B$ is said to be *normal* if there exists some constant $C$ such that for every $x, y \in X$ one has

$$\|x\| \leq C\|x + y\|$$

It is obvious that a normal cone is proper, namely $X \cap -X = \{0\}$.

To begin with, we indicate the relationships between the class of normal cones and the one of cones in the class $\mathcal{S}$; the weak uniform structure is defined after Proposition III.2.

## V.2 Proposition

Let $X \subset B$ be a convex cone and $\widehat{X}$ be its weak completion for the duality with $B'$.

1) If $X$ is a normal cone then $\widehat{X} \in \mathcal{S}$ and $\widehat{X} \subset B''$.

2) If $\widehat{X} \in \mathcal{S}$ then $X$ is a normal cone.

**Proof**

1) By Krein's Theorem (I.12) the polar cone $X^0$ of $X$ taken in $B'$ is closed and generating and by Klee's Theorem (I.15) every positive linear form on $B'$, for the preorder determined by $X^0$ is continuous, in other words $\widehat{X} \subset B''$. Since $X$ is assumed to be normal, by Krein's Theorem (I.12) every $\ell \in B'$ can be written $\ell = \ell_1 - \ell_2$ with $\ell_1, \ell_2 \in B'$ and $\ell_1, \ell_2$ positive on $X$; it follows that $\widehat{X}$ is proper, in other words $\widehat{X} \in \mathcal{S}$.

2) If $\widehat{X} \in \mathcal{S}$ by Lemma III.20 every $\ell \in B'$ can be written $\ell = \ell_1 - \ell_2$ with $\ell_1, \ell_2 \in B'$ and $\ell_1, \ell_2$ positive on $X$; it follows from Klee's Theorem (I.15) applied in $B'$ to $X^0$ that $X$ is a normal cone.

For continuous conical measures we refer to Definition IV.4.

## V.3 Proposition

If $X \subset B$ is a normal cone, every positive conical measure $\mu$ on $X$, for the duality with $B'$, is continuous.

Moreover there exists some constant $C$ such that:

$$\|r(\mu)\| \leq \|\mu\| \leq C\|r(\mu)\|$$

**Proof**

By the preceding proposition $\widehat{X} \subset B''$ and since $K_\mu \subset \widehat{X}$ it follows from Definition IV.4 that $\mu$ is continuous. Since $X$ is normal, by Krein's Theorem (I.12) there exists $C$ such that every $\ell \in B'$ can be written $\ell = \ell_1 - \ell_2$ with $\ell_1, \ell_2 \in B'$ and $\ell_1, \ell_2$ positive on $X$ and $\|\ell_1\| + \|\ell_2\| \leq C\|\ell\|$; since

$$\|\mu\| = \sup\{\mu(|\ell|) : \ell \in B_1'\}$$

and $\mu(|\ell|) \leq \mu(\ell_1 + \ell_2) = \ell_1(r(\mu)) + \ell_2(r(\mu))$ the righthandside inequality follows; the lefthandside inequality is obvious.

The following result, while simple, explains why it is possible to get further results for normal cones; for increasing norms we refer to Proposition I.11,5).

# V.4 Proposition

Let $X \subset B$ be a normal cone; there exists a constant $C$ such that, for every finite sequence $x_1,,,x_n$ in $X$, one has:

$$\|\sum_1^n x_i\| \leq M_1((x_i)) \leq C\|\sum_1^n x_i\|$$

When the norm of $B$ is increasing on $X$, we can take $C = 2$.

**Proof**

Since $M_1((x_i)) = \sup \|\sum_1^n \epsilon_i x_i\|$ over all the choice of signs, the lefthandside inequality follows; to prove the righthandside inequality, we can assume, by Proposition I.11, that the norm of $B$ is increasing on $X$: it suffices to note that

$$\|\sum_1^n \epsilon_i x_i\| \leq \|\sum_{k\in I} x_k\| + \|\sum_{k\in J} x_k\| \leq 2\|\sum_1^n x_i\|$$

where $k \in I$ iff $\epsilon_k = 1$ and $k \in J$ iff $\epsilon_k = -1$.

The following lemma of approximation will be useful when the cone involved is assumed to be normal.

# V.5 Lemma

Let $X \subset B$ be a convex cone, not necessarily normal, and $\widehat{X}$ be its closure in $B''$ for $\sigma(B'', B')$; for every finite sequence $x_1,,,x_n$ in $\widehat{X}$ there exists a directed set of sequences $x_1^\alpha,,,x_n^\alpha$ in $X$, $\alpha \in \mathcal{U}$, such that $x_i^\alpha \to x_i$ for $\sigma(B'', B')$ along $\mathcal{U}$, for every $1 \leq i \leq n$, and for every $\alpha \in \mathcal{U}$:

$$\|\sum x_i^\alpha\| \leq \|\sum x_i\|$$

**Proof**

We follows the same lines as in the proof of Lemma IV.8: let $K$ be the diagonal of $B'_1$ in $(B')^n$; we set

$$E = (X^0)^n + K \subset (B')^n$$

To finish the proof, we note that $E$ is closed for the duality with $B^n$ and we apply the Bipolar Theorem (I.8) as in the proof of Lemma IV.8.

We are now in a position to complete Theorem IV.10; the operator $f \mapsto \widehat{f}$, with respect to a cone in $\mathcal{S}$, is defined in Proposition III.23.

## V.6 Theorem

Let $X \subset B$ be a normal cone; for every $1 < p < \infty$ the following three properties are equivalent:

1) $p \in I_X$.

2) There exists $C > 0$ such that, for every finite sequence $\ell_1,,,\ell_n$ in $B'$, denoting by $f$ the map defined on $B''$ by

$$x \mapsto (\sum_1^n |\ell_i(x)|^p)^{1/p}$$

one has, for every $x \in \widehat{X}$: $\widehat{f}(x) \leq C\|x\|(\sum_1^n \|\ell_i\|^p)^{1/p}$, where $\widehat{f}$ is defined with respect to the cone $\widehat{X} \in \mathcal{S}$.

3) Same condition as in 2) for every $x \in X$.

**Proof**

1) $\Rightarrow$ 2): by condition c) of Theorem IV.10 and Proposition V.3, there exists $C > 0$ such that for every $x \in \widehat{X}$ and every positive conical measure $\mu$ on $X$ with $r(\mu) = x$ one has $\mu(f) \leq C\|x\|(\sum_1^n \|\ell_i\|^p)^{1/p}$; the proof follows from Proposition III.23.

2) $\Rightarrow$ 1): the arguments are the same as above.

Since 2) implies 3), it suffices to prove the converse: if $x \in \widehat{X}$, for every $\epsilon > 0$ there exists a finite sequence $x_1,,,x_n$ in $\widehat{X}$ with $x = \sum_1^n x_i$ and $\widehat{f}(x) \leq \sum_1^n f(x_i) + \epsilon$; with the notations of the above lemma, for a convenient $\alpha$ one has $\widehat{f}(x) \leq \sum_1^n f(x_i^\alpha) + 2\epsilon$; thus $\widehat{f}(x) \leq \widehat{f}(\sum_1^n x_i^\alpha) + 2\epsilon$ and so 3) implies 2) since $x_i^\alpha \in X$ and $\|\sum_1^n x_i^\alpha\| \leq \|x\|$.

## V.7 Remark

When $X$ is the positive cone of a Banach lattice $B$ then, with the notations of the above theorem, $\widehat{f}$ is an affine continuous function on $X$, namely $\widehat{f}$ is the restriction to $X$ of an element of $B'$. This element is called the *Krivine element* mentioned in II.10,5); it is denoted $(\sum_1^n |\ell_i|^p)^{1/p} = \widehat{f}$.

More generally, for every finite sequence $x_1,,,x_n$ in $X$ the element $(\sum_1^n x_i^p)^{1/p}$ exists; it is the supremum of the elements $\sum_1^n a_i x_i$ for all positive $a_1,,,a_n$ with $\sum_1^n a_i^{p'} = 1$.

When $B$ is a Banach lattice and when the conditions of the preceding theorem are satisfied, concerning the positive cone of $B$, we say that the norm is $p$-convex.

We proved in Proposition IV.18 that if $X$ and $Y$ are two convex cones in $B$, then $i(X+Y) \geq \sup\{i(X), i(Y)\}$ and in IV.22 that this inequality can be strict; in the framework of normal cones we have the following result.

## V.8 Proposition

Let $X, Y \subset B$ be two normal cones contained in a same normal cone; then

$$i(X+Y) = \sup\{i(X), i(Y)\}$$

**Proof**

Since $X, Y$ are contained in a same normal cone we can assume that the norm of $B$ is increasing on $X+Y$; if $q > \sup\{i(X), i(Y)\}$ one has by Theorem IV.10, for some constants $C_X$ and $C_Y$:

$(\sum_1^n \|x_i\|^q)^{1/q} \leq C_X \| \sum_1^n x_i \|$ for every finite sequence $x_1,,,x_n$ in $X$.

$(\sum_1^n \|y_i\|^q)^{1/q} \leq C_Y \| \sum_1^n y_i \|$ for every finite sequence $y_1,,,y_n$ in $Y$.

Consequently $(\sum_1^n \|x_i + y_i\|^q)^{1/q} \leq (C_X + C_Y) \| \sum_1^n (x_i + y_i) \|$, since the norm is increasing on $X+Y$; the result follows.

## V.9 Remark

We shall give, in Appendix 3, another proof of the preceding proposition which is more in the spirit of the theory of Integral Representation, and which relies directly on Theorem IV.10,c); namely we shall prove that $I(X+Y) = \inf\{I(X), I(Y)\}$.

Now, if $X \subset B$ is a normal cone with $i(X) > 1$, we shall show that $i(X)$ is the maximum of all $r \geq 1$ such that $B$ contains uniform copies of spaces $\ell_n^r$ whose basis vectors are contained in $X$. We shall use a very important theorem due to J.L. Krivine [35] as it is indicated at the end of section C of our Introduction.

To begin with, we recall the following definition due to J.L. Krivine [35]:

# V.10 Definition

Let $V$ be a normed vector lattice, that is a vector lattice equipped with a norm such that $(|x| \leq |y|)$ implies $(\|x\| \leq \|y\|)$; if $p \geq 1$ we say that $V$ is of *genre* $\leq p$ if there exists some constant $C$ such that, for every finite sequence $x_1,,,x_n$ in $V^+$, with $x_i \wedge x_j = 0$ for $i \neq j$, one has:

$$(\sum_1^n \|x_i\|^p)^{1/p} \leq C\|\sum_1^n x_i\|$$

In the preceding definition the expression "genre $\leq p$" should be viewed as a whole, because J.L. Krivine does not define the "genre" of $V$; however this definition makes sense since one can check that if $p \leq q$ and if $V$ is of genre $\leq p$ then $V$ is of genre $\leq q$.

Here is the theorem of J.L. Krivine [35] (given without proof) which will be quite useful for us; if $(a_n)$ is a sequence in $\mathbb{R}$ for every $p \geq 1$ we denote by $\|(a_n)\|_p$ the classical $\ell^p$ norm, namely $(\sum_n |a_n|^p)^{1/p}$.

# V.11 Theorem

Let $V$ be a normed vector lattice and $G(V)$ be the infimum of the set of all $p \geq 1$ such that $V$ is of genre $\leq p$. Let $K(V)$ be the supremum of the set of all $r \geq 1$ such that, for every integer $n$ and every $\epsilon > 0$ there exist norm one vectors $x_1,,,x_n$ in $V^+$, with $x_i \wedge x_j = 0$ for $i \neq j$, satisfying, for every scalar sequence $a_1,,,a_n$:

$$(1-\epsilon)\|(a_i)\|_r \leq \|\sum_1^n a_i x_i\| \leq (1+\epsilon)\|(a_i)\|_r$$

Then $K(V) = G(V)$.

In order to apply the preceding theorem we associate in the following key lemma a normed vector lattice to every normal cone.

## V.12 Lemma

Let $X \subset B$ be a normal cone on which the norm is increasing. Let $M_d(X)$ be the vector lattice consisting of all conical measures of the form $\sum_x a_x \epsilon_x$ where the sum is finite and $x \in X$. If $\mu \in M_d(X)$ the map $\mu \mapsto \|\mu\| = \|r(|\mu|)\|$ (the norm in $B$ of the resultant of $|\mu|$) is a norm and, equipped with this norm, $M_d(X)$ is a normed vector lattice.

**Proof**

If $\|\mu\| = 0$ then $r(|\mu|) = 0$; since $X$ is proper it follows that $|\mu| = 0$. Let us prove the triangular inequality: if $\lambda, \mu \in M_d(X)$ one has $|\lambda + \mu| \leq |\lambda| + |\mu|$, thus $r(|\lambda+\mu|) \leq r(|\lambda|) + r(|\mu|)$; the results follows since the norm is increasing on $X$.

## V.13 Remark

Conversely, if $\mu \mapsto \|\mu\|$ is a norm on $M_d(X)$, then the norm is increasing on $X$: indeed let $x, y \in X$ be located on two different rays; one has

$$\|\epsilon_x + \epsilon_y\| = \|x + y\| \text{ and } \|\epsilon_x - \epsilon_y\| = \|x + y\|$$

If the triangular inequality holds true one has:

$$2\|x\| = \|(\epsilon_x + \epsilon_y) + (\epsilon_x - \epsilon_y)\| \leq 2\|x + y\|$$

which means that the norm is increasing on $X$.

The following lemma will enable us to apply Theorem V.11 to the space $M_d(X)$.

## V.14 Lemma

One has $i(X) = G(M_d(X))$.

**Proof**

Let $p > G(M_d(X))$; if $x_1,,,x_n \in X$ are located on $n$ different rays, then $\epsilon_{x_1},,,\epsilon_{x_n}$ are disjoint elements of $M_d(X)$; thus, since $\|\sum_1^n \epsilon_{x_i}\| = \|\sum_1^n x_i\|$, one has for some constant $G_p$:

$$(\sum_1^n \|x_i\|^p)^{1/p} \leq G_p \|\sum_1^n x_i\|$$

This means that $p \geq i(X)$; the case when some elements are located on the same ray is easy.

Let $p > i(X)$; if $\mu_1,,,\mu_n \in M_d(X)$ are positive, not necessarily disjoint, one has $\| \sum_1^n \mu_i\| = \|r(\sum_1^n \mu_i)\|$; it follows that for some constant $S_p$ one has:

$$(\sum_1^n \|\mu_i\|^p)^{1/p} \leq S_p\| \sum_1^n \mu_i\|$$

This means that $p \geq G(M_d(X))$.

Here is the first result we are aiming for:

## V.15 Theorem

Let $X \subset B$ be a normal cone with $i(X) > 1$; if the norm is increasing on $X$ then $i(X)$ is the greatest $r \geq 1$ for which, for every integer $n$ and every $\epsilon > 0$, there exist norm 1 elements $x_1,,,x_n \in X$ such that, for every $a_1,,,a_n \geq 0$, one has:

$$(1-\epsilon)\|(a_i)\|_r \leq \| \sum_1^n a_i x_i\| \leq (1+\epsilon)\|(a_i)\|_r$$

**Proof**

We set $p = i(X)$; by Krivine's Theorem (V.11) and Lemma V.12, there exist $\mu_1,,\mu_n \in M_d(X)$, positive and of norm 1, such that

$$(1-\epsilon)\|(a_i)\|_p \leq \| \sum_1^n a_i \mu_i\| \leq (1+\epsilon)\|(a_i)\|_p$$

Since $\|\mu_i\| = \|r(\mu_i)\|$ and $\| \sum_1^n a_i\mu_i\| = \| \sum_1^n a_i r(\mu_i)\|$ it suffices to take $x_i = r(\mu_i)$ for $1 \leq i \leq n$.

Conversely, suppose that $r > i(X)$ satisfies the conclusion of the theorem we would have

$$\| \sum_1^n x_i\| \leq 2n^{1/r}$$

but if $i(X) < q < r$ we have $n^{1/q} \leq C_q\| \sum_1^n x_i\|$; this is a contradiction when $n \to \infty$.

## V.16 Remark

When $i(X) = 1$ the preceding theorem remains true but it has no interest since, in this case, we can take $x_1 = x_2 = .. = x_n$.

From Theorem V.15 we can deduce the following result.

# V.17 Theorem

Let $X \subset B$ be a normal cone with $i(X) > 1$. Then $i(X)$ is the greatest $r$ for which there exist $0 < m \leq M < \infty$ such that, for every integer $n$, there exist norm one elements $x_1,,,x_n \in X$ such that, for every scalars $a_1,,,a_n$, one has:

$$m\|(a_i)\|_r \leq \|\sum_1^n a_i x_i\| \leq M\|(a_i)\|_r$$

Moreover, when the norm is increasing on $X$, for every $\epsilon > 0$, one can take:

$$m = 2^{1-1/i(X)} - 1 - \epsilon \text{ and } M = 2^{1-1/i(X)} + \epsilon$$

**Proof**

To show that, if $r > i(X)$, then $r$ cannot have the required properties it suffices to mimic the beginning of the proof of Theorem V.14.

In the rest of the proof we assume that the norm is increasing on $X$ and we follow the notations of Theorem V.15, with $r = i(X)$:

Let $a_1,,,a_n \in \mathbb{R}$ with $\|(a_i)\|_r = 1$; we set $\|\sum_1^n a_i x_i\| = \delta$ and $u = \sum_1^n a_i^+ x_i$, $v = \sum_1^n a_i^- x_i$. One has:

$$(1-\epsilon)^r \sum_1^n (a_i^+)^r \leq \|u\|^r \leq (1+\epsilon)^r \sum_1^n (a_i^+)^r$$

$$(1-\epsilon)^r \sum_1^n (a_i^-)^r \leq \|v\|^r \leq (1+\epsilon)^r \sum_1^n (a_i^-)^r$$

Let us prove the assertion concerning $m$:

Since $2u = (u+v) + (u-v)$ and $2v = (u+v) + (v-u)$, one has:

$$2\|u\| \leq 1 + \delta \text{ and } 2\|v\| \leq 1 + \delta$$

Since $\|(a_i)\|_r = 1$ it follows that $\|u\|^r + \|v\|^r \geq (1-\epsilon)^r$ and that

$$2^r(1-\epsilon)^r \leq 2(1+\delta)^r$$

letting $\epsilon \to 0$ we get $\delta \geq 2^{1-1/r} - 1$, which is the desired result.

Let us prove the assertion concerning $M$:

One has $\delta = \|u - v\| \leq \|u\| + \|v\|$ and consequently

$$\delta/2 \leq (\|u\|^r/2 + \|v\|^r/2)^{1/r}$$

Thus $(\delta/2)^r \leq (1+\epsilon)^r/2$; letting $\epsilon \to 0$ we get $\delta \leq 2^{1-1/r}$, which is the desired result.

# V.18 Remark

To prove the preceding theorem it is important to have at our disposal in Theorem V.16 the numbers $(1-\epsilon)$ and $(1+\epsilon)$, namely numbers arbitrary close to 1; in fact we can prove the following result:

Let $x_1,,,x_n$ be a norm 1 sequence in $B$ such that, for every sequence of positive scalars $a_1,,,a_n$, one has:

$$a\|(a_i)\|_p \leq \|\sum_1^n a_i x_i\| \leq A\|(a_i)\|_p$$

where $0 < a \leq 1 \leq A < \infty$ and $p \geq 1$. If $A < (2^{1-1/p})a$ there exist $0 < m \leq M < \infty$ such that for every sequence of scalars one has:

$$m\|(a_i)\|_p \leq \|\sum_1^n a_i x_i\| \leq M\|(a_i)\|_p$$

The proof is the same as the proof of the preceding theorem.

Theorem V.17 enables us to get some inequalities involving $i(X)$, the type index $p_B$ and the cotype index $q_B$, for which we refer to Appendix 1.

# V.19 Theorem

Let $X \subset B$ be a normal cone; if $i(X) > 1$ one has $p_B \leq i(X) \leq q_B$.

**Proof**

The righthandside inequality holds true for every convex cone as we have seen in Proposition IV.17. The lefthandside inequality follows from Theorem V.17 and the following result due to B. Maurey, G. Pisier and J.L. Krivine [46]:

$p_B$ is the smallest $r \geq 1$ such that $B$ contains uniform copies of $\ell_n^r$.

# V.20 Theorem

Let $X$ be a normal cone contained in a Hilbert space $H$; then either $i(X) = 1$ or $i(X) = 2$.

**Proof**

We apply the preceding theorem, noticing that for a Hilbert space $H$ one has: $p_H = q_H = 2$.

## V.21 Remark

Let $B = L^p([0,1], dx)$, with $1 \leq p \leq 2$ and $X = L^p_+$; we have proved in IV.21 that $i(X) = p$; consequently, by Theorem V.19 for each convex cone $Y \subset X$ one has $i(Y) = 1$ or $i(Y) = p$.

More generally the preceding property holds for every normal cone $X \subset B$ such that $i(X) = p_B$.

## V.22 Remark

When the convex cone $X \subset B$ is not normal, we have no result analogous to Theorem V.15 or to Theorem V.17. However, notice that, in the case when $X = B$, then the result holds true, by a Theorem of B. Maurey, G. Pisier and J.L. Krivine [46]. Consequently there are two opposite cases: when $X$ is small (normal) and when $X$ is the whole space.

## V.23 Problem

In the case when $X \subset B$ is a normal cone, Theorem V.17 establishes the existence of a greatest number, namely $i(X)$, having some properties; we have no formula giving a minimun or an infimum in this framewok. However, in the case when $X = B$ this minimum does exist; it is nothing but the type index $p_B$ of $B$, by a Theorem of B. Maurey, G. Pisier and J.L. Krivine [46].

In the framework of Theorem V.15 the minimum is obviously 1; it suffices to take $x_1 = \cdots = x_n$.

We shall now deal with the case when the cone $X \subset B$ is closed and both normal and generating ($B = X - X$); it follows then from Krein's Theorem (I.13) and Klee's Theorem (I.15) that in $B'$ the polar cone $X^0$ of $X$ is also both normal and generating. The following lemma will be useful.

## V.24 Lemma

Let $\Gamma$ be a (non-empty) convex subset of a vector space $V$ (without any topology). Let $f : \Gamma \to \mathbb{R}$ be a concave function and $g : \Gamma \to \mathbb{R}$ be a convex function with $f \leq g$ on $\Gamma$. Then, there exists an affine function $\ell : \Gamma \to \mathbb{R}$, such that $f \leq \ell \leq g$ on $\Gamma$.

**Proof**

First, we deal with the case when $\Gamma$ is finite dimensional; we may assume that the interior of $\Gamma$, denoted by $\Gamma_0$, is non-empty. For every integer $n$, by the Separation Theorem I.4 there exists an affine function $\ell_n$ on $\Gamma_0$ such that $f \leq \ell_n \leq g+1/n$ on $\Gamma_0$ and thus also on $\Gamma$ (it suffices to check on every segment in $\Gamma$). If $\mathcal{U}$ is any non-trivial ultrafilter on $\mathbb{N}$ then $\ell = \lim_{\mathcal{U}}$ is convenient.

In the general case, we note that $\Gamma$ is the (directed) union of its finite dimensional convex subsets $F$; let $\mathcal{F}$ denote the family of those $F$ and for every $F \in \mathcal{F}$ let $\ell_F$ be an affine function on $F$ such that $f \leq \ell_F \leq g$ on $F$; it suffices to take $\ell = \lim(\ell_F)$ along some convenient ultrafilter on $\mathcal{F}$.

We are now in a position to give a characterization of the set $I_X$ in terms of the order determined on $B'$ by the polar cone $X^0$, when $X$ is closed, normal and generating.

## V.25 Theorem

Let $X \subset B$ be a closed convex cone, normal and generating. For every $p > 1$ the following two properties are equivalent:

1) $p \in I_X$.

2) There exists $C > 0$ such that, for every finite sequence $\ell_1,,,\ell_n$ in $B'$, there exists $\ell \in B'$ satisfying, for every scalar sequence $a_1,,,a_n$ with $\|(a_i)\|_{p'} = 1$:

$$\sum_1^n a_i \ell_i \leq \ell$$

for the order on $B'$ determined by $X^0$.

Moreover $\ell$ satisfies: $\|\ell\| \leq C(\sum_1^n \|\ell_i\|^p)^{1/p}$.

**Proof**

1) $\Rightarrow$ 2): by Theorem V.6, with its notations, for every $x \in X$, one has

$$\widehat{f}(x) \leq C\|x\|(\sum_1^n \|\ell_i\|^p)^{1/p}$$

We apply Lemma V.24 to the two functions $x \mapsto \widehat{f}(x)$ and $x \mapsto C\|x\|(\sum_1^n \|\ell_i\|^p)^{1/p}$ defined on $X$, to get an affine function $\varphi$ on $X$ such that, for every $x \in X$, one has:

$$\widehat{f}(x) \leq \varphi(x) \leq C\|x\|(\sum_1^n \|\ell_i\|^p)^{1/p}$$

Since $X$ is generating, $\varphi$ can be extended obviously to $\ell \in X^0$ with the required control of $\|\ell\|$.

Since $f(x) = \sup\{\sum_1^n a_i \ell_i(x) : \|(a_i)\|_{p'} = 1\}$ for every $x \in X$, it is clear that $\ell$ has the required majorization properties.

2) $\Rightarrow$ 1): we see that $\widehat{f}(x) \leq \ell(x)$ for every $x \in X$; the result follows from Theorem V.6.

We shall now study a class of operators on $B$ (see Definition V.26), equipped with a normal cone. For this purpose we shall use some results concerning the following two classes of operators, involving a Banach lattice $L$ and a Banach space $E$:

1) For $1 \leq p \leq q$, the class of $(q,p)$-*concave* operators from $L$ to $E$.

2) For $1 \leq p \leq q$, the class of $(p,q)$-*convex* operators from $E$ to $L$.

For the definition of these two classes of operators we refer to Appendix 2 where some results are mentioned.

Here is the class of operators we will study:

## V.26 Definition

Let $Y$ be a convex cone contained in a Banach space $E$ and $F$ be another Banach space. A linear operator $T : E \to F$ is said to be $(q,1)$-*summing on* $Y$, where $q \geq 1$, if there exists $C > 0$ such that, for every finite sequence $x_1, , , x_n$ in $Y$ one has:

$$(\sum_1^n \|T(x_i)\|^q)^{1/q} \leq C M_1((x_i))$$

The following lemma will enables us to use some results mentioned in Appendix 2.

## V.27 Lemma

Let $X \subset B$ be a normal cone on which the norm is increasing and $\rho : M_d(X) \to B$ be the map $\rho(\mu) = r(\mu)$. For every continuous linear map $u$ from $B$ to a Banach space $F$ the following two properties are equivalent:

1) $u$ is $(q,1)$-summing on $X$.

2) $u \circ \rho$ is $(q,1)$-concave.

**Proof**

1) $\Rightarrow$ 2): let $\mu_i = \sum_x a_x^i \epsilon_x$ be a finite family of elements of $M_d(X)$, where each $\sum$ is finite and $x \in X$ satisfies $\|x\| = 1$; one has:

$$(\sum_i \|u \circ \rho(\mu_i)\|^q)^{1/q} = \big(\sum_i \|\sum_x a_x^i u(x)\|^q\big)^{1/q} \leq$$

$$\leq \big(\sum_i \|\sum_x (a_x^i)^+ u(x)\|^q\big)^{1/q} + \big(\sum_i \|\sum_x (a_x^i)^- u(x)\|^q\big)^{1/q} \leq$$

$$\leq 2C\|\sum_{i,x} |a_x^i|x\| = 2C\|\sum_i |\mu_i|\|$$

because the norm of $B$ is increasing on $X$; thus $u \circ \rho$ is $(q, 1)$-concave.

2) $\Rightarrow$ 1): if $x_1,,,x_n \in X$ it suffices to consider $\epsilon_1,,,\epsilon_n \in M_d(X)$ to get the result.

To get further results, we need to determine the dual space of $M_d(X)$; recall that a superlinear function on a convex cone is a positively homogeneous concave function.

## V.28 Proposition

Let $X \subset B$ be a normal cone on which the norm is increasing. The continuous dual space of $M_d(X)$ can be identified with the vector space of all positively homogeneous functions $f : X \to \mathbb{R}$ such that $|f|$ is majorized by a superlinear function, which is bounded on $X \cap B_1$. One has:

$$\|f\| = \sup\{\sum_1^n |f(x_i)| : x_1,,,x_n \in X \text{ and } \|\sum_1^n x_i\| \leq 1\}$$

If $\mu = \sum_x a_x \epsilon_x$, where the sum is finite and $x \in X$, then $\mu(f) = \sum_x f(x)$.

The transposed map $\rho' : B' \to (M_d(X))'$ associates to every $\ell \in B'$ the restriction of $\ell$ to $X$.

**Proof**

Let $T \in (M_d(X))'$; for $x \in X$, we set $f(x) = T(\epsilon_x)$; it is clear that $f$ is homogeneous; if $\mu = \sum_x a_x \epsilon_x$ then $T(\mu) = \sum_x a_x f(x)$. Since $M_d(X)$ is a normed lattice, one has:

$$\|T\| = \sup\{\sum_1^n |f(x_i)| : x_1,,x_n \in X \text{ and } \|\sum_1^n x_i\| \leq 1\}$$

Now, it suffices to note that the map defined on $X$ by

$$x \mapsto \sup\{\sum_1^n |f(x_i)| : x_1,,,x_n \in X \text{ and } \sum_1^n x_i = x\}$$

is the smallest function on $X$ which is superlinear and greater than $|f|$.

Conversely, if $f$ is homogeneous on $X$ and is majorized by some superlinear function $g$ on $X$, which is bounded on $X \cap B_1$ then:

$$|\mu(f)| = |\sum_x a_x f(x)| \leq \sum_x |a_x||f(x)| \leq \sum_x |a_x| g(x) \leq g(\sum_x |a_x| x)$$

because g is superlinear; since $g$ is bounded on $X \cap B_1$ and $\|\mu\| = \|\sum_x |a_x| x\|$, it follows that $f$ is an element of $(M_d(X))'$.

We are now in a position to prove the result we are aiming for:

## V.29 Theorem

Let $X \subset B$ be a closed normal and generating cone. If $u$ is a continuous linear operator from $B$ to another Banach space $F$, the following three properties are equivalent for $q > 1$:

1) $u$ is $(q, 1)$-summing on $X$.

2) For every $1 < p < q$, there exists a constant $C_p$ such that for every finite sequence $\ell_1,,,\ell_n$ in $F'$ there exists $\ell \in X^0$ satisfying:

a) For every scalar sequence $a_1,,,a_n$ with $\|(a_i)\|_p = 1$, one has, for the order determined on $B'$ by $X^0$:

$$\sum_1^n a_i u'(\ell_i) \leq \ell$$

b) $\|\ell\| \leq C_p(\sum_1^n \|\ell_i\|^{q'})^{1/q'}$.

3) Same condition as 2) for some $1 < p < q$.

**Proof**

By Proposition I.11, we can assume that the norm is increasing on $X$.

1) $\Rightarrow$ 2): by lemma V.27, the map $u \circ \rho$ is $(q, 1)$-concave and it is also $(q, p)$-concave for every $1 < p < q$ by a theorem of B. Maurey [43] recalled in Appendix 2. By a result recalled in Appendix 2, $(u \circ \rho)'$ is $(q', p')$-convex; since $(u \circ \rho)' = \rho' \circ u'$ one has:

$$\|(\sum_1^n |\rho' \circ u'(\ell_i)|^{p'})^{1/p'}\| \leq C_p(\sum_1^n \|\ell_i\|^{q'})^{1/q'}$$

Consequently, by Proposition V.28, there exists a superlinear function $\phi$ : $X \to \mathbb{R}$ such that for every $x \in X$ one has:

$$(\sum_1^n |(u'(\ell_i))(x)|^{p'})^{1/p'} \leq \phi(x) \leq C_p\|x\|(\sum_1^n \|\ell_i\|^{q'})^{1/q'}$$

We complete the proof, as in Theorem V.25, applying Lemma V.24 to the two functions $x \mapsto \phi(x)$ and $x \mapsto C_p\|x\|(\sum_1^n \|\ell_i\|^{q'})^{1/q'}$

Since 2) $\Rightarrow$ 1) is obvious, we prove that 3) $\Rightarrow$ 1): for every $x \in X$ we see that:

$$(\sum_1^n |(u'(\ell_i))(x)|^{p'})^{1/p'} \leq \ell(x)$$

thus

$$\|(\sum_1^n |\rho' \circ u'(\ell_i)|^{p'})^{1/p'}\| \leq C_p(\sum_1^n \|\ell_i\|^{q'})^{1/q'}$$

and $(u \circ \rho)'$ is $(q', p')$-convex. By a Theorem recalled in Appendix 2, $u \circ \rho$ is $(q, p)$-concave, thus $(q, 1)$-concave by a Theorem of B. Maurey [43], recalled in Appendix 2; finally $u$ is $(q, 1)$-summing on $X$ by Lemma V.27.

## V.30 Remark

When $X \subset B$ is normal (and not necessarily closed and generating) it is clear that condition 2 of Theorem V.6 is equivalent to the fact that the canonical map from $B'$ to $(M_d(X))'$ is $(p, p)$-convex ($p$-convex for short).

## V.31 Remark

When $X \subset B$ is not normal, we can define the space $M_d(X)$ as before, namely as the vector space of all $\mu = \sum_x a_x \epsilon_x$, where the sum is finite and $x \in X$; we set $\|\mu\| = M_1((a_x x))$.

It is clear that, when $X$ is normal and the norm of $B$ is increasing on $X$, the norm just defined is equivalent to the one defined in Lemma V.12. But, in the general case, we cannot get results analogous to Theorem V.15 or V.17, because, if $\mu_1, , , \mu_n \in M_d((X))$ are positive and pairwise disjoint, it is not sure that $\| \sum_1^n a_i \mu_i \| = \| \sum_1^n a_i r(\mu_i) \|$ for positive scalars $a_1, , , a_n$.

## V.32 Notes and remarks

1) The proof given by J.L. Krivine [35] of his Theorem (V.11) relies partly on some calculations involving the growth of convenient series in the positive cone of the given normed vector lattice. When I tried to prove Theorem V.15, I first tried to mimic these calculations in the framewok of normal cones; but it seemed to me that it was quite intricated and I decided

to look for a device which would enable me to apply directly Krivine's Theorem, and this was the construction of the space $M_d(X)$ given in Lemma V.12.

Some years after H. Lemberg [40] gave a somewhat simpler presentation of Krivine's proof, described in [45] by B. Maurey.

Notice that Krivine's Theorem is no longer stated in the original form given by Krivine in [35], as we did in V.11, but in a somewhat different form so as to fit with the proof of the Theorem of B. Maurey, G. Pisier and J.L. Krivine [46] recalled in Appendix 1.

For an extended discussion of these results we refer to B. Maurey's Article [45].

2) Theorem V.2 is taken from ([14], VIII.1.7); Theorem V.6 is taken from [7], [9]; V.8 is new; for Theorem V.15 and the space $M_d(X)$ see [8]; Theorem V.17 is proved (in a somewhat particular case) in [10] and in the general case in [11] along with the inequalities V.19 and V.20. For the remainder of the Chapter see [10].

# Chapter VI

# Positive operators defined on $\mathcal{C}(K)$ or with values in $L^1$

There are two parts in this Chapter. In the first one we study operators from a space $\mathcal{C}(K)$ to a Banach space $B$ equipped with a convex cone $X$ (not necessarily proper), which are positive in the sense that $\mathcal{C}^+(K)$ is mapped into $X$ (Theorem VI.1). When $X = B$ we retrieve a Theorem due to B. Maurey [42].

Next, we give applications to the localization of conical measures (Theorems VI.7, 8), using the interpretation of conical measures due to E. Thomas mentioned in Proposition III.18.

In the second part of this Chapter we study operators from a Banach space $B$, equipped with a convex cone $X$, not necessarily generating, to a space $L^1(P)$, where $P$ is a probability, which are positive in the sense that $X$ is mapped into $L^1_+(P)$ (Theorem VI.12). When $B$ is a subspace of $L^1(P)$, and $X = 0$, we retrieve a Theorem due to H.P. Rosenthal [59] (Theorem VI.13).

Finally, we give applications to the study of closed convex subcones of $L^1_+(P)$.

As in Chapter IV, we shall use only very basic results concerning summing operators; Chapter VII is devoted to these operators.

## VI.1 Theorem

Let $X \subset B$ be a convex cone (non necessarily proper) such that $i(X) < +\infty$ and let $r > i(X)$. There exists a $C > 0$ such that, for every Hausdorff compact space $K$, if $T : \mathcal{C}(K) \to B$ is a positive continuous operator (in the sense that $T(f) \in X$ whenever $f \geq 0$), then $T$ is $r$-summing and $\pi_r(T) \leq C\|T\|$.

Conversely, if $r > 1$ satisfies the preceding properties, then $r \geq i(X)$.

**Proof**

We shall prove that there exists a Radon probability measure $P$ on $K$ such that, for every $f \in \mathcal{C}(K)$, one has $\|T(f)\| \leq C(\int |f(u)|^r P(du))^{1/r}$ and the result follows easily from Pietsch's Inequality (Theorem VII.4).

Let $T' : B' \to \mathcal{M}(K)$ be the adjoint map of $T$; it is easy to see that $T'$ is a positive map for the order determined on $B'$ by the polar cone $X^0$ of $X$. It is a classical fact that $\mathcal{M}(K)$ is isometric to a space $L^1(\Omega, \mathcal{A}, m)$ for some measure space $(\Omega, \mathcal{A}, m)$: indeed, if $(m_\alpha)$ is a maximal set of Radon probabilities on $K$ with $m_\alpha \wedge m_\beta = 0$ for $\alpha \neq \beta$, then $\mathcal{M}(K)$ is nothing but the $\ell^1$ direct sum $(\oplus_\alpha L^1(m_\alpha))_1$; consequently, $\mathcal{M}(K) = L^1(\Omega, \mathcal{A}, m)$ ($L^1(m)$ for short) for some convenient measure space $(\Omega, \mathcal{A}, m)$.

Since $r > i(X)$, by Theorem IV.15, $T'$ admits a factorization through the space $L^{r'}(m)$; so there exists some positive norm one function $g \in L^r(m)$ such that:

1) For every $\ell \in B'$ one has $\{T'(\ell) \neq 0\} \subset \{g > 0\}$
2) For some constant $C_r$, for every $\ell \in B'$ one has:

$$\|T'(\ell)/g\|_{r'} \leq C_r \|T'\| \|\ell\|$$

Condition 1) implies that there exists some Radon probability $m_0$ on $K$ such that, for every $\ell \in B'$ one has $T'(\ell) \in L^1(m_0)$; consequently, by abuse of notation, $g$ can be viewed as a norm one element of $L^r(m_0)$.

For every $f \in \mathcal{C}(K)$, there exists $\ell \in B'_1$ such that:

$$\|T(f)\| = \langle T(f), \ell \rangle = \langle f, T'(\ell) \rangle = \int_K f(\omega) T'(\ell)(\omega) m_0(d\omega)$$

$$\int_K f(\omega) g(\omega) (T'(\ell)(\omega)/g(\omega)) m_0(d\omega)$$

From Holder's Inequality we get:

$$\|T(f)\| \leq C_r \|T\| (\int_K |f(\omega)|^r g(\omega)^r m_0(d\omega))^{1/r}$$

The result follows since $g(\omega)^r m_0(d\omega)$ is a Radon probability measure $P$ on $K$.

Conversely, let $r \geq 1$ and $C > 0$ such that, for every compact Hausdorff space $K$, every positive continuous map $T : \mathcal{C}(K) \to B$ is $r$-summing and satisfies $\pi_r(T) \leq C\|T\|$:

Let $x_1, , , x_n$ be a finite sequence in $X$ and let $K = \{1, , , n\}$; the map $T : \mathcal{C}(K) \to B$ defined by $T(a_1, , , a_n) = \sum_1^n a_i x_i$ is positive and satisfies $\|T\| = M_1((x_i))$; for $1 \leq i \leq n$ let $f_i : K \to \mathbb{R}$ defined by $f_i(j) = 1$ if $j = i$ and 0 otherwise; one has $T(f_i) = x_i$ and $M_r((f_1, , , f_n)) = 1$.

From the hypothesis it follows that:

$$(\sum_1^n \|T(x_i)\|^r)^{1/r} \leq CM_1((x_i))$$

Consequently $r \geq i(X)$.

## VI.2 Remark

When the cone $X$ is the whole space $B$, the preceding theorem is due to B. Maurey [42]; in fact, it was stated in terms of cotype taking into account that $q_B = i(B)$. Moreover, when $B$ is of cotype 2 the preceding theorem holds true with $r = 2$; this result is due to E. Dubinsky, A. Pelczynski and H.P. Rosenthal [21].

Consequently, when $B$ is a $L^p$ space, with $1 \leq p \leq 2$ we retrieve a weaker result than the classical one.

When $B$ is a $L^p$ space, with $2 < p < \infty$, we retrieve a result due to H.P. Rosenthal [59].

We shall apply Theorem VI.1 to a basic problem concerning the theory of conical measures; using the notations of Chapter III let us introduce the following

## VI.3 Definition

Let $E$ be a Hausdorff l.c.s.. A conical measure $\mu \geq 0$ on $E$ is said to be *localizable* if there exists a Radon measure $m \geq 0$ on $E \setminus \{0\}$ such that:

1) Every $\ell \in E'$ is $m$-integrable.

2) For every $f \in h(E)$ one has $\mu(f) = m(f)$.

We say that $m$ is a *localization* of $\mu$ on $E \setminus \{0\}$.

## VI.4 Remark

1) If $\mu \neq 0$ such a localization is never unique because for every $x \neq 0$ and every $k > 0$ we have: $\epsilon_x = k\epsilon_y$ with $y = x/k$.

2) Conversely, if $m \geq 0$ is a Radon measure on $E \setminus \{0\}$ then $m$ can be viewed as a localization of some conical measure $\mu \geq 0$ on $E$ iff every $\ell \in E'$ is $m$-integrable.

In the framework of the Theory of Integral Representation it is important to deal with localizable conical measures because, if $\mu \geq 0$ is localizable by the Radon measure $m \geq 0$ on $E \setminus \{0\}$ and if $m$ is carried by a subset $K \subset E \setminus \{0\}$ consisting of "interesting" elements, then $r(\mu)$ can be viewed as a "mixture" of the elements of $K$.

For more details we refer to our book [14].

We recall the definition of *nuclear operators*:

## VI.5 Definition

If $E$ and $F$ are two Banach spaces, a continuous linear map $T : E \to F$ is said to be *nuclear* if there exist a sequence $(\ell_n)$ in $E'$ and a sequence $(y_n)$ in $F$ with $\sum \|\ell_n\| \|y_n\| < +\infty$ such that, for every $x \in E$:

$$T(x) = \sum \ell_n(x) y_n$$

The following lemma is due to E. Thomas [66]; we use the notations of Proposition III.16.

## VI.6 Lemma

Let $K$ be a compact Hausdorff space and $\Lambda$ be the canonical conical measure on $\mathcal{C}(K)$. If $T$ is a nuclear map from $\mathcal{C}(K)$ to a Banach space $E$ then $T(\Lambda)$ is localizable on $E \setminus \{0\}$.

**Proof**

It is well known that if $\sum a_n$ is a convergent series of positive terms, there exists a sequence $b_n \to +\infty$ such that $\sum b_n a_n < +\infty$.

Consequently, there exist a sequence $(\mu_n)$ of Radon measures on $K$ with $\sum \|\mu_n\| < +\infty$ and a sequence $(y_n)$ in $E$ with $y_n \to 0$ such that, for every $x \in \mathcal{C}(K)$, one has:

$$T(x) = \sum_n \mu_n(x) y_n$$

By Radon-Nikodym's Theorem there exist a Radon probability measure $m$ on $K$ and a sequence $(f_n)$ in $L^1(m)$ with $\sum_n |f_n| \in L^1(m)$ such that, for every $x \in \mathcal{C}(K)$, one has:

$$T(x) = \sum_n \left( \int x(t) f_n(t) m(dt) \right) y_n$$

There exists a sequence $(A_k)$ of pairwise disjoint compact subsets of $K$ with $m(\cup A_k) = 1$ such that for every $k$ the sum $\sum_i |f_i|$ and every $f_i$ are continuous on $A_k$; consequently, $m$-a.e., the series $\rho(t) = \sum_n f_n(t)y_n$ converges in $E$. For every $k$ it is easy to see that $\rho$ is continuous on $A_k$ since $y_n \to 0$. It follows that for every $k$ the image under $\rho$ of the restriction of $m$ to $A_k$ is a finite Radon measure on $E$.

The image $m_0 = \rho(m)$ is a Radon measure on $E \setminus \{0\}$: as a matter of fact every $\ell \in E'$ is $m_0$-integrable because

$$\rho(m_0)(|\ell|) = \int |\ell(\rho(t)|m(dt) \leq \sum_n \int |f_n(t)||\ell(y_n)|m(dt) < +\infty$$

(the last inequality holds true since $T$ is nuclear).

For every $\ell \in E'$ one can check that $T'(\ell)$ is the Radon measure on $K$ with density (with respect to $m$) the function $t \mapsto \ell(\rho(t))$, which is an element of $L^1(m)$.

For a finite sequence $\ell_1,,,\ell_n$ in $E'$, we define $f \in h(E)$ by $f(x) = \sup(\ell_1(x),,, \ell_n(x))$; by definition:

$$T(\Lambda)(f) = \sup\{T'(\ell_1),,,T'(\ell_n)\}(1)$$

and it follows that $T(\Lambda)(f) = \int f(\rho(t))m(dt)$; in other words $m_0 = \rho(m)$ is a localization of $T(\Lambda)$ on $E \setminus \{0\}$.

We are now in a position to prove the two results concerning the localization of conical measures we are aiming for.

## VI.7 Theorem

Let $X \subset B$ be a convex cone with $i(X) < 2$ and let $\mu \geq 0$ be a conical measure on $B$ such that $K_\mu \subset B$. If $E$ is a Banach space and $T : B \to E$ is a 2-summing linear operator, then $T(\mu)$ is localizable on $E \setminus \{0\}$.

**Proof**

With the notations of Proposition III.18, since $K_\mu \subset X$ there exist a compact Hausdorff space $K$ and a map $\varphi : \mathcal{C}(K) \to B$, which is positive in the sense that $\varphi(f) \in X$ whenever $f \geq 0$, such that $\varphi(\Lambda) = \mu$.

One has $T(\mu) = (T \circ \varphi)(\mu)$; by Theorem VI.1, $\varphi$ is 2-summing and by a classical Theorem of A. Grothendieck [26], [27] or ([20], 5.31) $T \circ \varphi$ is nuclear; the result follows from Lemma VI.6.

# VI.8 Theorem

Let $\mu \geq 0$ be a conical measure on $B$ with $K_\mu \subset B$; if $B$ has cotype 2, then for every Banach space $E$ and every 2-summing map $T : B \to E$ then $T(\mu)$ is localizable on $E \setminus \{0\}$.

**Proof**

We mimic the proof of the preceding theorem using the fact (Remark VI.2) that $\varphi$ is 2-summing since $B$ has cotype 2.

# VI.9 Applications

If $\mu \geq 0$ is a conical measure on the positive cone $X$ of an $L^p$ space where $1 < p < 2$ we can apply Theorem VI.7 because $X \in \mathcal{S}$ and so $K_\mu \subset X$. For $p = 2$ we can apply Theorem VI.8.

We deal now with positive linear operators from a Banach space $B$, ordered by a closed convex cone $X$ (not necessarily proper), with values in a space $L^1(P)$, where $P$ is a probability.

We need the following classical lemma, which implies Radon-Nikodym's Theorem.

# VI.10 Lemma

Let $(\Omega, \mathcal{A}, P)$ be a probability space; every positive linear form $m$ on $L^\infty(\Omega, \mathcal{A}, P)$ admits a unique decomposition $m = \varphi dP + m_s$ where $\varphi \in L^1_+(P)$ and $m_s \geq 0$ (the singular part of $m$) is finitely additive on $\mathcal{A}$ and satisfies that for some $g \in L^\infty(\Omega, \mathcal{A}, P)$ with $g > 0$ (a.e.) one has $m(g) = 0$

**Proof**

Let $\varphi$ be the greatest $f \in L^1(P)$ such that $fdP \leq m$, in the sense of linear form on $L^\infty(\Omega, \mathcal{A}, p)$; it is easy to see that such a $\varphi$ exists; we set $m_s = m - \varphi dP$. Consequently, for every set $A \in \mathcal{A}$, with $P(A) > 0$, and every $\epsilon > 0$, it is not true that $\epsilon 1_A dP \leq m_s$, thus there exists an $\mathcal{A}$-measurable set $A_\epsilon \subset A$ with $m_s(A_\epsilon) \leq \epsilon P(A_\epsilon)$ and $P(A_\epsilon) > 0$.

For every $n > 0$ we can construct a sequence $E(n,i)$ $(i \in \mathbb{N})$ of $\mathcal{A}$-measurable pairwise disjoint sets and an integer $k_n$ such that:

1) For every $n$ one has: $P(\cup_i E(n,i)) = 1$.

2) For every $n$ and every $i$ one has: $m_s(E(n,i)) \leq P(E(n,i))/n$.

3) For every $n$ one has: $P(\cup_{k_n}^\infty E(n,i)) \leq 2^{-n}$.

For every $n$ define $g_n : \Omega \to \mathbb{R}$ by:

a) $g_n(\omega) = 1$ on every set $E(n,i)$ for $1 \leq i < k_n$.

b) $g_n(\omega) = 1/n$ on every set $E(n,i)$ for $i \geq k_n$.

Then, for every $n$ one has $m_s(g_n) \leq 2/n$; now it suffices to set $g = \inf_n(g_n)$: one has $m_s(g) = 0$ and $g$ satisfies $g > 0$ a.e. by Borel-Cantelli's Lemma, since the sets $E_n = \cup_{k_n}^{\infty} E(n,i)$ satisfies $P(\cup E_n) \leq \sum 2^{-n} < +\infty$.

## VI.11 Remark

The preceding lemma implies Radon-Nikodym's Theorem: indeed, if $m$ is $\sigma$-additive then $m_s$ is also $\sigma$-additive and since $m_s(g) = 0$ it follows that $m_s((g > 0)) = 0$ and it suffices to notice that $P((g > 0)) = 1$.

## VI.12 Theorem

Let $X \subset B$ be a convex cone such that $\overline{X} \neq B$ and $i(X^0) < +\infty$. For every $r > i(X^0)$ there exists some constant $C_r$ such that every continuous positive map $T : B \to L^1(\Omega, \mathcal{A}, P)$, where $P$ is a probability, admits a factorization through $L^{r'}(P)$ in the following sense:

There exists $f_T \in L^r(P)$ with $f_T > 0$ (a.e.) and $\|f_T\|_r = 1$, such that for every $x \in B$

$$\|T(x)/f_T\|_r \leq C_r \|T\| \|x\|$$

Conversely, if $r$ satisfies the preceding properties, then $r \geq i(X^0)$.

**Proof**

A) In the direct sense, there are two steps.

1) It is well-known that $L^\infty(\Omega, \mathcal{A}, P)$ ($L^\infty(P)$ for short) is linearly isometrically isomorphic to a space $\mathcal{C}(K)$ for some compact Hausdorff space $K$; consequently, it follows from Theorem VI.1 that there exists a constant $C_r$ such that, since $T'$ is positive when $B'$ is equipped with $X^0$, then $T'$ is $r$-summing.

By Pietsch's Inequality (Theorem VII.4) there exists a positive linear form $m \geq 0$ on $L^\infty(P)$ such that for every $f \in L^\infty(P)$ one has:

$$\|T'(f)\| \leq (m(|f|^r))^{1/r}$$

with $m(1) \leq C_r^r \|T\|^r$.

With the notations of Lemma VI.10, if we denote by $I(g)$ the ideal of $L^\infty(P)$ generated by $g$, for every $f \in I(g)$ one has:

$$\|T'(f)\| \leq (\int |f(\omega)|^r \varphi(\omega) P(d\omega))^{1/r}$$

2) We set $\psi = \varphi^{1/(1-r)}$ and for every $s \geq 1$ we denote by $||| \, |||_s$ the norm in $L^s(\Omega, \mathcal{A}, \psi dP)$.

a) If $x \in B$ satisfies $|||T(x)|||_{r'} < +\infty$ $(1/r + 1/r' = 1)$, then for every $\epsilon > 0$ there exists $g_\epsilon \in I(g)$, with $|||g_\epsilon|||_r \leq 1$, such that:

$$|||T(x)|||_{r'} \leq \int T(x)(\omega) g_\epsilon(\omega)\psi(\omega)P(d\omega) + \epsilon$$

There exists $\psi_\epsilon \in L^\infty(P)$, with $|\psi_\epsilon| \leq \psi$, such that, with obvious abuse of notation:

$$|||T(x)|||_{r'} \leq \int T(x)(\omega) g_\epsilon(\omega)\psi_\epsilon(\omega)P(d\omega) + 2\epsilon = \langle x, T'(g_\epsilon \psi_\epsilon)\rangle + 2\epsilon$$

thus

$$|||T(x)|||_{r'} \leq \|x\| \|T'(g_\epsilon \psi_\epsilon)\| + 2\epsilon$$

Since $g_\epsilon \psi_\epsilon \in I(g)$ and $|\psi_\epsilon| \leq \psi$, using the last inequality of step 1), it follows that:

$$|||T(x)|||_{r'} \leq \|x\| (\int |g_\epsilon(\omega)|^r |\psi(\omega)|^r \varphi(\omega)P(d\omega))^{1/r} + 2\epsilon \leq \|x\| + 2\epsilon$$

Thus

$$(\int |T(x)|^{r'}(\omega)\varphi^{1/(1-r)}(\omega)P(d\omega))^{1/r} \leq \|x\|$$

If we set: $f_T = \varphi^{1/r}/(\int \varphi dP)^{1/r}$, obviously $\|f_T\|_r = 1$ and, since

$$\int \varphi dP \leq m(1) \leq C_r^r \|T\|^r$$

we get:

$$(\int |T(x)/f_T|^{r'} dP)^{1/r'} \leq C_r \|T\| \|x\|$$

b) If $x \in B$ satisfies $|||T(x)|||_{r'} = +\infty$, using the same path as in a) it is easy to derive a contradiction.

B) Conversely, let $r > 1$ satisfy the hypothesis; if $\ell_1,,, \ell_n$ is a finite sequence in $X^0$, define $T : B \to \ell_n^1$ by $T(x) = (\ell_1(x),,, \ell_n(x))$; one has $\|T\| = M_1((\ell_i))$.

By the hypothesis there exists a finite positive sequence $f_T = (f_1,,, f_n)$ of norm 1 in $\ell_n^r$ such that:

$$\|T(x)/f_T\|_{r'} \leq C_r \|T\| \|x\|$$

In particular, for every $1 \leq i \leq n$, one has: $\|\ell_i\| \leq C_r f_i \|T\|$; it follows that:

$$(\sum \|\ell_i\|^r)^{1/r} \leq C_r M_1((\ell_i))$$

thus $r \geq i(X^0)$, which is the result we are aiming for.

We are now in a position to retrieve a classical theorem of H.P. Rosenthal [59].

## VI.13 Theorem

Let $B \subset L^1(P)$ be a closed subspace of infinite dimension; if $B$ is reflexive, then $B$ is isomorphic to a closed subspace of some space $L^p(Q)$ for some probability $Q$. More precisely one has $2 \leq i(B') < \infty$ and for every $r > i(B')$ there exist $f_0 \in L^{r'}(P)$ of norm 1 with $f_0 > 0$ a.e. and $C$ such that, for every $f \in B$ one has:

$$\|f\|_1 \leq \|f/f_0\|_{r'} \leq C\|f\|_1$$

**Proof**

By a Theorem of B. Maurey and G. Pisier [46] (recalled in Appendix 1) $i(B')$ is nothing but the cotype index $q_{B'}$ of $B'$, thus $i(B') \geq 2$.

1) If $i(B') < \infty$, for every $r > i(B')$ we apply Theorem VI.1 with $X = 0$; there exist $f_0 \in L^r(P)$ of norm 1 with $f_0 > 0$ a.e. and $C$ such that for every $f \in B$:

$$\|f\|_1 \leq \|f/f_0\|_{r'} \leq C\|f\|_1$$

2) If $i(B') = +\infty$, by a classical Theorem of B. Maurey, G. Pisier and J.L. Krivine [46] (recalled in Appendix 1) $B$ contains uniform copies of $\ell_n^\infty$ and so $B$ is not reflexive.

We shall apply the preceding results to the study of convex cones contained in $L^1(P)$, where $P$ stands for a probability.

## VI.14 Definition

If $X \subset L^1(P)$ is a convex cone and if $1 < p < \infty$, we say that $X$ has property $F_p$ if there exist $C$ and $f_0 \in L^{p'}(P)$ of norm 1 with $f_0 > 0$ a.e. such that for every $f \in X$:

$$\|f/f_0\|_p \leq C\|f\|_1$$

Notice that by Holder's Inequality one has: $\|f\|_1 \leq \|f/f_0\|_p$.

# VI.15 Lemma

Let $X \subset L^1(P)$ be a convex cone; for $1 < p < \infty$ the following three properties are equivalent:

1) $X$ has property $F_p$.

2)The closure $\overline{X}$ of $X$ in $L^1(P)$) has property $F_p$.

3) The convex cone generated by all the functions $|f|$, where $f \in X$ has property $F_p$.

**Proof**

1) $\Rightarrow$ 2): let $(f_n)$ be a sequence in $X$ such that $f_n \to f_0$ in $L^1(P)$ and $f_n \to f_0$ a.e.; since $\|f_n/f_0\|_p \leq C\|f_n\|_1$ it suffices to apply Fatou's Lemma.

2) $\Rightarrow$ 1) is obvious.

1) $\Rightarrow$ 3): let $f_1, f_2 \in X$; one has:

$$\|(|f_1| + |f_2|)/f_0\|_p \leq \|f_1/f_0\|_p + \|f_2/f_0\|_p \leq C\||f_1|\|_1 + C\||f_2|\|_1$$

the result follows since the norm is additive on the positive cone of $L^1(P)$.

3) $\Rightarrow$ 1) is obvious.

# VI.16 Examples

1) For $1 < p < \infty$ and $f_0 \in L^{p'}(P)$ of norm 1 with $f_0 > 0$ a.e. and $C > 1$ the set $X$ of all $f \in L^1_+(P)$ such that $\|f/f_0\|_p \leq C\|f\|_1$ is a closed convex cone having property $F_p$; the proof is the same as the one of Lemma VI.15.

   Notice that the set $\{f : f \in X, \ \|f\|_1 \leq 1\}$ is relatively compact in $X$, by Dunford-Pettis's Theorem: indeed, for every $f \in X$, one has:

$$\int_e f dP = \int 1(e) f_0 (f/f_0) dP \leq C\|f\|_1 (\int_e f_0^{p'} dP)^{1/p'}$$

   and the hypothesis of Dunford-Pettis's Theorem are satisfied.

2) Here is an explicit example ([20], 1.10): let $(r_n)$ be the sequence of Rademacher's functions on $[0,1]$ equipped with $P = dx$; by Khintchine's inequalities for every $1 < p < \infty$ there exist $A_p, B_p$ such that for every finite sum $f = \sum_1^n a_i r_i$ one has $A_p\|f\|_1 \leq \|f\|_p \leq B_p\|f\|_1$. Consequently, the vector space generated by Rademacher's functions has propery $F_p$ in $L^1(P)$.

   Notice that the set of all the elements $|f|$ is a cone in $L^1(dx)$ which is not necessarily convex.

Now, we shall study the class of closed convex cones contained in the positive cone of a space $L^1(P)$.

# VI.17 Theorem

Let $X$ be a closed convex subcone of $L^1_+(P)$; if we equip $B = X - X$ with the norm

$$\|f\|_B = \inf\{\|f_1\|_1 + \|f_2\|_1 : f = f_1 - f_2\ f_1, f_2 \in X\}$$

then:

1) $B$ is a Banach space and $\|f\|_B = \|f\|_1$ for $f \in X$; moreover $X$ is closed, normal and generating in $B$. If $\ell \in X^0$ ($X^0$ taken in $B'$) then $\|\ell\|_{B'} = \sup\{\ell(f) : f \in X,\ \|f\|_1 \leq 1\}$.

2) If $i(X^0) < \infty$ ($X^0$ taken in $B'$), for every $r > i(X^0)$ there exist $f_r \in L^r(P)$ with $f_r > 0$ a.e. and $C_r > 0$ such that:

$$\|f/f_r\|_{r'} \leq C_r\|f\|_1 \text{ for every } f \in X$$

3) The following two properties are equivalent:

a) $i(X^0) = \infty$ ($X^0$ taken in $B'$).

b) For every $\epsilon > 0$ and every integer $n$ there exist $\ell_1,,,\ell_n \in X^0$, of norm 1, such that $\sum_1^n \ell_i(f) \leq (1+\epsilon)\|f\|_1$ for every $f \in X$.

**Proof**

1) Since $X$ is closed in $L^1(P)$ it is clear that $B$ is a Banach space; it is clear that $\|\ \|_B$ and $\|\ \|_1$ coincide on $X$ and that $X$ is generating and normal, since the norm is additive on $X$.

2) Apply Theorem VI.12.

3) a) $\Rightarrow$ b): apply Theorem V.15.

b) $\Rightarrow$ a): if $i(X^0) < \infty$, there exist $1 < p < \infty$ and $C$ such that $\|\sum_1^n \ell_i\| \geq Cn^{1/p}$ which contradicts b).

# VI.18 Notes and remarks

1) Most of the results of this Chapter come from [12] where some useless restrictions were involved (normality of $X$ in Theorem VI.1 and $X$ generating in Theorem VI.12). When $X = B$ Theorem VI.1 is due to B. Maurey [42]; his proof ([42] Proposition 3) is different of the one given here; it is also the case for the proof given by B. Maurey and G. Pisier ([46] Proposition 1.4). The proof given in ([20], 11.14.b) uses two Theorems of interpolation of G. Pisier ([20],10.8, 10.9).

   We emphasize that the proof given here uses the result that $i(B) = q_B$.

2) Lemma VI.6 is due to E. Thomas [66] in a different framework: roughly speaking, let $X$ be a closed convex cone contained in a Hausdorff l.c.s. $E$. If $\mu \geq 0$ is a conical measure on $E$ with $K_\mu \subset X$ the set $K_1 = K_\mu - K_\mu$

is symmetric and $\sigma(E, E')$ compact. Let us assume a condition of *co-nuclearity*, namely that there exists a symmetric compact set $K_2 \subset E$, with $K_1 \subset K_2$, such that, if $E_1$ and $E_2$ are the two Banach spaces whose unit balls are $K_1$ and $K_2$, the indentity map $\varphi : E_1 \to E_2$ is nuclear:

It follows from Lemma VI.6 that $\mu$ is localizable on $E$; one can prove that $\mu$ is localizable on $X$.

3) Lemma VI.10 is attributed by S.R. Foguel [23] to J. Neveu who told me, about 30 years ago, that he was quite surprised by this attribution. We emphasize that our proof implies Radon-Nikodym's Theorem.

4) Concerning theorem VI.13 due to H.P. Rosenthal [59], let us indicate again that his Article is one of the three roots of B. Maurey's work [44], the others (see II.10) are an Article by E.M. Nikishin and a theorem from A. Grothendieck's Résumé.

   Let us describe briefly the path followed by H.P. Rosenthal: first (in his Proposition 3) he studies, for $1 \leq p < \infty$, the Banach spaces $B$ such that every operator from a space $\mathcal{C}(K)$ to $B$ is $p$-summing. Then (in his Theorem 1) he applies this result to the case when $B$ is a subspace of a space $L^1(P)$: namely, he investigates when the canonical quotient map from $L^\infty$ onto $B'$ is $p$-summing. Notice that, in the proof of his Theorem 1, H.P. Rosenthal does not use the factorization Theorem stated in Chapter II.7.

# Chapter VII

# Summing operators on cones and integral operators

In Chapters IV and VI we have met summing operators on convex cones: in IV.10 we used mainly the definition of these operators; in IV.13 and VI.1 we used Pietsch's Inequality (Theorem VII.4). In this Chapter we shall study these operators more deeply. We prove Pietsch's Inequality in VII.4, we prove a Composition Theorem in VII.11 and we investigate transposed maps in VII.17. These results will be useful in Chapter VIII. When no cone is involved the results contained in this Chapter are classical. Finally, we describe some (classical) results concerning *integral operators* in VII.27, which will be useful in Chapter VIII.

For convenience we recall the definition of summing operators on a cone, given in the Introduction (Definition 6).

## VII.1 Definition

Let $X \subset B$ be a convex cone (not necessarily proper) and $p \geq 1$; if $T$ is a linear operator (not necessarily continuous), from $B$ to another Banach space, $T$ is said to be *$p$-summing on $X$* if there exists some $C > 0$ such that, for every finite sequence $x_1, , , x_n$ in $X$ one has:

$$(\sum_{1}^{n} \|T(x_i)\|^p)^{1/p} \leq C M_p((x_i))$$

We denote by $\pi_p(X, T)$ the smallest possible constant $C$.

If $E$ is a Banach space, we denote by $\Pi_p(X; B, E)$ the set of all (not necessarily continuous) linear operators $T : B \to E$ which are $p$-summing on $X$.

If $X = B$ we write for short $\pi_p(T)$ and $\Pi_p(B, E)$.

## VII.2 Examples

We have encountered a lot of examples of such operators in the preceding Chapters:

1) In Lemma IV.9 we have seen that, if $B$ is a Banach space, then for every finite sequence $\ell_1,,, \ell_n$ in $B'$ the map $T : B \to \ell_n^p$ defined by $x \mapsto (\ell_i(x))$ is $p$-summing on $B$ and $\pi_p(T) \leq (\sum_1^n \|\ell_i\|^p)^{1/p}$.

2) In Chapter VI we investigated the case when $B$ is a $\mathcal{C}(K)$ space.

3) For every compact Hausdorff space $K$, every Radon probability measure $P$ on $K$ and $1 \leq p < \infty$, the identity map $T : \mathcal{C}(K) \to L^p(P)$ is $p$-summing on $\mathcal{C}(K)$ and $\pi_p(T) = 1$: the fact that $\pi_p(T) \leq 1$ is easy to check from definition VII.1; the fact that $\pi_p(T) = 1$ follows from Theorem VII.4.

4) With the notations of Definition VII.1, if $T$ is $p$-summing on $X$, for every continuous linear map $U$ from $E$ to a Banach space, then $U \circ T$ is also $p$-summing on $X$.

For every continuous linear map $V$, from a Banach space $F$ to $B$, if $T$ is $p$-summing on $X$, then $T \circ V$ is $p$-summing on $V^{-1}(X)$.

When the convex cone $X \subset B$ is big enough we have the following.

## VII.3 Proposition

Let $X \subset B$ be a closed convex generating cone and $E$ be another Banach space; for every $p \geq 1$ then every $T \in \Pi_p(X; B, E)$ is continuous and $\pi_p(X, \ )$ is a norm on $\Pi_p(X; B, E)$, which is a Banach space when it is equipped with this norm.

**Proof**

For evey $x \in X$ one has $\|T(x)\| \leq \pi_p(X, T)\|x\|$; since $X$ is closed and generating it follows from Klee's Theorem (I.15) that $T$ is continuous and that $\|T\| \leq C\pi_p(X, T)$ for some constant $C$. Let $(T_n)$ be a Cauchy sequence in $\Pi_p(X; B, E)$; by the preceding inequality $(T_n)$ converges in the norm topology to a limit denoted $T$.

For every $\epsilon > 0$, there exists $n_\epsilon$ such that, for every $n, m > n_\epsilon$ and every finite sequence $x_1,,, x_k$ in $X$ one has:

$$(\sum_1^k \|(T_n - T_m)(x_i)\|^p)^{1/p} \leq \epsilon M_p((x_i))$$

If we fix $\epsilon > 0$ and $m$ and if we let $n \to \infty$, it follows that $T \in \Pi_p(X; B, E)$; it is easy to see that $T_n \to T$ in $\Pi_p(X; B, E)$.

The following result, called Pietsch's Inequality, is fundamental.

## VII.4 Theorem

Let $X \subset B$ be a convex cone and $E$ be another Banach space; if $T$ is a (not necessarily continuous) linear operator from $B$ to $E$, for every $p \geq 1$ the following two properties are equivalent:

1) $T$ is $p$-summing on $X$.

2) There exists a Radon measure (called a Pietsch measure) $m \geq 0$ on $B'_1$ equipped with $\sigma(B', B)$ such that, for every $x \in X$ one has:

$$\|T(x)\| \leq \left(\int |\ell(x)|^p m(d\ell)\right)^{1/p}$$

Moreover, $\pi_p(X, T)$ is nothing but the infimum of $m(1)^{1/p}$ for all such measures.

3) Same condition as 2) where $B'_1$ is replaced by any closed set $K \subset B'_1$ which is a norming set, namely for every $x \in B$ there exists $\ell \in K$ such that $\ell(x) = \|x\|$.

When $X$ is a normal cone the above conditions are equivalent to the following:

4) There exists a Radon measure $m \geq 0$ on $X^0 \cap B'_1$ such that, for every $x \in X$ one has:

$$\|T(x)\| \leq \left(\int (\ell(x))^p m(d\ell)\right)^{1/p}$$

Moreover, $\pi_p(X, T)$ is equivalent to the infimum of $(m(1))^{1/p}$ for all such measures.

**Proof**

1) $\Rightarrow$ 2): we shall apply Lemma II.4. Let $\mathcal{F}$ be the family of functions defined on $B'_1$ of the following form:

To every finite sequence $s = (x_1, , , x_n)$ in $X$ we associate the function $f_s$ defined on $B'_1$ by:

$$f_s(\ell) = \pi_p(X, T)^p \left(\sum_1^n |\ell(x_i)|^p\right) - \left(\sum_1^n \|T(x_i)\|^p\right)$$

$\mathcal{F}$ is obviously a cone of continuous functions, when $B'_1$ is equipped with $\sigma(B', B)$; this cone is convex because, if $f_{s_1}, f_{s_2} \in \mathcal{F}$, then $f_{s_1} + f_{s_2}$ is nothing but $f_s$ where $s$ is the finite sequence in $X$ obtained by gluing together $s_1$ and $s_2$.

By the hypothesis every function $f_s$ takes on a positive value on $B'_1$; thus, by Lemma II.4, there exists a Radon probability $P$ on $B'_1$ such that $\int f(\ell)P(d\ell) \geq 0$ for every $f \in \mathcal{F}$. Consequently if $s = \{x\}$, where $x \in X$, we get:

$$\|T(x)\| \leq \pi_p(X,T)(\int |\ell(x)|^p P(d\ell))^{1/p}$$

2) $\Rightarrow$ 1): for $x_1,,,x_n$ in $X$ we have:

$$\sum_1^n \|T(x_i)\|^p \leq \int (\sum_1^n |\ell(x_i)|^p) m(d\ell) \leq m(1) M_p((x_i))^p$$

The assertion concerning $\pi_p(X,T)$ follows from the preceding calculation.
1) $\Rightarrow$ 3): see the proof of 1) $\Rightarrow$ 2), noticing that

$$M_p((x_i)) = \sup\{\ell(\sum_1^n a_i x_i)\} : \ell \in K, \|(a_i)\|_{p'} \leq 1\}$$

When $X$ is a normal cone let us prove that 1) implies 4): it suffices to mimic the proof that 1) implies 2) noticing that, since $X$ is normal, by Krein's Theorem (I.13) there exists $C$ such that, for every finite sequence $x_1,,,x_n$ in $X$ one has:

$$M_p((x_i)) \leq C \sup\{\sum_1^n \ell(x_i)^p : \ell \in X^0 \cap B'_1\}$$

The proof that 4) implies 1) is the same as for 2) implies 1).

From the preceding theorem we deduce:

## VII.5 Proposition

For $1 \leq p \leq q < \infty$ one has: $\Pi_p(X;B,E) \subset \Pi_q(X;B,E)$ and for every $T \in \Pi_p(X;B,E)$ one has: $\pi_q(X,T) \leq \pi_p(X,T)$.

When $X$ is closed and generating and if $\Pi_p(X;B,E) = \Pi_q(X;B,E)$, then the two norms $\pi_p(X,)$ and $\pi_q(X,\ )$ are equivalent.

**Proof**

The first assertion follows from the classical inequality:

$$(\int f^p(u)P(du))^{1/p} \leq (\int f^q(u)P(du))^{1/q}$$

which holds true for every function $f \geq 0$ and every probability $P$.

For the second assertion, note that the identity map from $\Pi_p(X; B, E)$ onto $\Pi_q(X; B, E)$ is continuous from the preceding assertion; consequently, by the open mapping theorem, it is an isomorphism.

When no cone is involved, namely when $X = B$, we retrieve the following classical Pietsch's Factorization Theorem [54].

## VII.6 Theorem

Let $B$ and $E$ be two Banach spaces and $1 \leq p < \infty$; for every continuous linear $T : B \to E$ the following two properties are equivalent:

1) $T$ is $p$-summing on $B$.

2) There exists some Radon measure $m \geq 0$ on $B'_1$, equipped with $\sigma(B', B)$, such that, if $i_B$ denotes the canonical embedding from $B$ to $\mathcal{C}(B'_1)$, if $i_p$ denotes the canonical embedding from $\mathcal{C}(B'_1)$ to $L^p(m)$, and if $F$ is the closure in $L^p(m)$ of the range of $i_p \circ i_B$ there exists a continuous linear map $T_p : F \to E$ such that:

$$T = T_p \circ i_p \circ i_B$$

Moreover, $\pi_p(T)$ is the infimum of the set of all products $m(1)^{1/p} \|T_p\|$.

3) Same condition as 2) (mutatis mutandis) when $B'_1$ is replaced by some closed norming set $K \subset B'_1$ (for every $x \in B$ there exists $\ell \in K$ such that $\ell(x) = \|x\|$).

**Proof**

1) $\Rightarrow$ 2): we apply Pietsch's Inequality (Theorem VII.4); it suffices to note that if $x \in B$ satisfies $\int |i_B(x)(u)|^p m(du) = 0$ then $T(x) = 0$; consequently there exists a continuous map $T_p$ defined on the range of $i_p \circ i_B$ such that $T = T_p \circ i_p \circ i_B$, and $T_p$ extends to $F$ by continuity.

2) $\Rightarrow$ 1): this is easy using Examples VII.2,3).

Condition 3) is proved in the same way.

When $X$ is a normal cone and $p = 1$ we have the following:

## VII.7 Proposition

Let $X \subset B$ be a normal cone; if $T$ is a (not necessarily continuous) linear map from $B$ to another Banach space $E$ the following two properties are equivalent:

1) $T$ is 1-summing on $X$.

2) There exists $\ell_T \in X^0$ such that one has: $\|T(x)\| \leq \ell_T(x)$ for every $x \in X$.

Moreover $\pi_1(X,T)$ is equivalent to the infimum of $\|\ell_T\|$ taken over all the convenient $\ell_T \in X^0$.

**Proof**

1) $\Rightarrow$ 2): by Theorem VII.4 there exist a Radon probability $P$ on $X^0 \cap B'_1$ and $C$ such that for every $x \in X$ one has $\|T(x)\| \leq C \int \ell(x) P(d\ell)$. It suffices to choose $\ell_T = C\ell_P$ where $\ell_P$ is the barycenter of $P$ which exists (by Lemma II.3) and belongs to $X^0 \cap B'_1$ since this set is convex and compact for $\sigma(B', B)$.

2) $\Rightarrow$ 1) is obvious and $\pi_1(X,T) \leq \|\ell_T\|$.

When $X$ is closed, normal and generating and $p = 1$, we have the following factorization Theorem due to B. Walsh ([67], p. 151).

## VII.8 Theorem

Let $X \subset B$ be a closed normal and generating cone. If $E$ is another Banach space, then for every $T : B \to E$ the following properties are equivalent:

1) $T$ is 1-summing on $X$.

2) There exists $L \in X^0$ such that $\|T(x)\| \leq L(x)$ for every $x \in X$.

3) There exist a Banach space $\widehat{B}$, ordered by a normal and generating cone $\widehat{X}$, such that the norm is additive on $\widehat{X}$, a positive continuous map $\Theta : B \to \widehat{B}$ and a continuous map $\widehat{T} : \widehat{B} \to E$ such that $T = \widehat{T} \circ \Theta$.

4) There exists $L \in X^0$ such that the set $T'(E'_1)$ is contained in the order interval $]-L, L[$ for the order on $B'$ determined by the polar cone $X^0$.

**Proof**

1) $\Rightarrow$ 2) is a case of Theorem VII.7.

2) $\Rightarrow$ 3): we equip $B$ with the following semi-norm:

$$p(x) = \inf\{L(x_1 + x_2) : x = x_1 - x_2,\ x_1, x_2 \in X\}$$

We define $\widehat{B}$ as the Hausdorff completion of $B$ and $\widehat{X}$ is the closure of $X$ in $\widehat{B}$; it is easy to see that $p(x) = L(x)$ for every $x \in X$, thus $\widehat{X}$ is a normal cone in $\widehat{B}$; since every $x \in B$ with $p(x) < 1$ can be written $x = x_1 - x_2$ with $p(x_1) + p(x_2) < 1$, it follows that $\widehat{X}$ is generating in $\widehat{B}$.

Let $\Theta$ be the canonical map from $B$ to $\widehat{B}$; since $X$ is generating in $B$, by Klee's Theorem (I.15) there exists $C$ such that every $x \in B$ can be written $x = x_1 - x_2$, with $\|x_1\| + \|x_2\| \leq C\|x\|$; it follows that

$$p(x) \leq L(x_1) + L(x_2) \leq \|L\|(\|x_1\| + \|x_2\|) \leq \|L\|C\|x\|$$

thus $\Theta$ is continuous.

For every $x \in B$, if $x = x_1 - x_2$ with $x_1, x_2 \in X$, one has:

$$\|T(x)\| \leq \|T(x_1)\| + \|T(x_2)\| \leq L(x_1 + x_2)$$

thus $\|T(x)\| \leq p(x)$ and it is easy to finish the proof.

3) $\Rightarrow$ 1): for every finite sequence $x_1,,, x_n$ in $X$ one has

$$\sum_1^n \|T(x_i)\| \leq \|\widehat{T}\|(\sum_1^n \|\Theta(x_i)\|) = \|\widehat{T}\|\|\Theta(\sum_1^n x_i)\| \leq \|\widehat{T}\|\|\Theta\|\|\sum_1^n x_i\|$$

and the result follows.

2) $\Rightarrow$ 4): for every $\ell \in B'_1$ and every $x \in X$ one has: $-L(x) \leq \ell(T(x)) \leq L(x)$; the righthandside inequality yields $L(x) - T'(\ell)(x) \geq 0$, thus $L - T'(\ell) \in X^0$ and it is easy to finish the proof.

4) $\Rightarrow$ 2): for every $x \in X$, there exists $\ell \in B'_1$ with $\|T(x)\| = \ell(T(x)) = T'(\ell)(x)$; from condition 4) it follows that $\|T(x)\| \leq L(x)$.

Now we shall prove a "Composition Theorem" concerning summing operators on convex cones.

The following classical lemma will help us to deal with convex cones:

# VII.9 Lemma

Let $1 \leq p < \infty$; for every finite sequence $x_1,,, x_n$ in $B$ one has:

$$M_p((x_i)) \leq 2 \sup\{\|\sum_1^n a_i x_i\| : a_1,,, a_n \geq 0, \ \|(a_i)\|_{p'} \leq 1\}$$

**Proof**

Let $\epsilon > 0$; with an error of at most $\epsilon$ there exist $\ell \in B'_1$ and a sequence $a_1,,, a_n$ of real numbers with $\|(a_i)\|_{p'} \leq 1$ such that:

$$M_p((x_i)) = \sum_1^n a_i \ell(x_i) \leq \sum_+ |a_i| \ell(x_i) - \sum_- |a_i| \ell(x_i)$$

where $\sum_+$ concerns the set $\{i : \ell(x_i) > 0\}$ and $\sum_-$ concerns the set $\{i : \ell(x_i) < 0\}$; the result follows.

The following lemma is the main step towards the result we are aiming for (Pietsch measures are defined in VII.4).

# VII.10 Lemma

Let $E$ and $F$ be two Banach spaces, $X \subset E$ be a convex cone (not necessarily proper) and $v : E \to F$ be $q$-summing on $X$, for some $q > 1$.

For $1 \leq r < q$, define $p$ by $1/r = 1/p + 1/q$. Then for every finite sequence $x_1,,,x_n$ in $X$ with $M_r((x_i)) \leq 1$ there exist $\sigma_1,,,\sigma_n \geq 0$ and $y_1,,,y_n$ in $F$ such that:

1) $v(x_i) = \sigma_i.y_i$ for $i = 1,,,n$.

2) $\|(\sigma_i)\|_q \leq m(1)^{1/q}$, where $m$ is any Pietsch measure for $v$ on $B_1'$.

3) $M_p((y_i)) \leq 2$.

Moreover, if $X = E$ condition 3) can be replaced by:

4) $M_p((y_i)) \leq 1$.

**Proof**

For every $x \in X$, we set $\sigma(x) = (\int |\ell(x)|^r m(d\ell))^{1/q}$ and we denote $\sigma(x_i)$ by $\sigma_i$ for short. Then $\|(\sigma_i)\|_q \leq m(1)^{1/q}$ and we get 2).

We assume that $\sigma_i > 0$ for every $i = 1,,,n$ and we set $y_i = v(x_i)/\sigma_i$ to get 1).

It remains to prove 3). For every $a_1,,,a_n \geq 0$ with $\|(a_i)\|_{p'} \leq 1$ one has:

$$\|\sum_1^n a_i.y_i\| = \|\sum_1^n a_i.\sigma_i^{-1}.v(x_i)\| = \|v(\sum_1^n a_i.\sigma_i^{-1}.x_i)\|$$

Since $a_1,,,a_n \geq 0$ and $\sigma_1,,,\sigma_n \geq 0$ one has $(\sum_1^n a_i.\sigma_i^{-1}.x_i) \in X$ and we can apply Pietsch's Inequality (VII.4) to get:

$$\|\sum_1^n a_i.y_i\| \leq (\int |\ell(\sum_1^n a_i.\sigma_i^{-1}.x_i)|^q m(d\ell))^{1/q}$$

and

$$\|\sum_1^n a_i.y_i\| \leq (\int (\sum_1^n a_i.\sigma_i^{-1}.|\ell(x_i)|)^q m(d\ell))^{1/q}$$

We can write $\sum_1^n a_i.\sigma_i^{-1}.|\ell(x_i)| = \sum_1^n A_i.B_i.C_i$, where

$$A_i = a_i^{p'/q}.\sigma_i^{-1}.|\ell(x_i)|^{r/q}, B_i = |\ell(x_i)|^{1-r/q}, C_i = a_i^{1-p'/q}$$

We shall apply the following general form of Holder's Inequality (for positive elements):

$$\sum_i U_i.V_i.W_i \leq (\sum_i U_i^\alpha)^{1/\alpha}(\sum_i V_i^\beta)^{1/\beta}(\sum_i W_i^\gamma)^{1/\gamma}$$

which holds true whenever $\alpha, \beta, \gamma > 0$ satisfy $1/\alpha + 1/\beta + 1/\gamma = 1$.

We can choose $\alpha = q$, $\beta = p$, $\gamma = r'$ because $1/q + 1/p + 1/r' = 1$. Since $p(1 - r/q) = r$ and $r'(1 - p'/q) = p'$ we get:

$$\sum_1^n A_i.B_i.C_i \leq (\sum_1^n a_i^{p'}.\sigma_i^{-q}|\ell(x_i)|^r)^{1/q}(\sum_1^n |\ell(x_i)|^r)^{1/p}(\sum_1^n a_i^{p'})^{1/r'}$$

Since $M_r((x_i)) \leq 1$ and $\|(a_i)\|_{p'} \leq 1$ it follows that:

$$\sum_1^n A_i.B_i.C_i \leq (\sum_1^n a_i^{p'}.\sigma_i^{-q}|\ell(x_i)|^r)^{1/q}$$

Since $\| \sum_1^n a_i.y_i\| \leq (\int(\sum_1^n A_i.B_i.C_i)^q m(d\ell))^{1/q}$, by the formula giving $\sigma_i$ we get:

$$\|\sum_1^n a_i.y_i\| \leq (\int \sum_1^n a_i^{p'}.\sigma_i^{-q}.|\ell(x_i)|^r)m(d\ell))^{1/q} = (\sum_1^n a_i^{p'})^{1/q} = 1$$

which is condition 3).

When $X = E$, we imitate the above proof; indeed the proof is simpler, we note that $M_p((x_i)) = \sup\{\| \sum a_i x_i\| : \|(a_i)\|_{p'} \leq 1\}$ and we drop Lemma VII.9.

When the operators are summing on the Banach spaces involved, the following "Composition Theorem" is due to A. Pietsch [54].

## VII.11 Theorem

If $T_1 : E \to F$ and $T_2 : F \to G$ are two operators, such that $T_1$ is $p_1$-summing on a convex cone $X \subset E$ and $T_2$ is $p_2$-summing on a convex cone $Y \subset F$ with $T_1(X) \subset Y$, then $T_2 \circ T_1$ is $r$-summing on $X$, with $1/r = \inf\{1, 1/p_1 + 1/p_2\}$.

Moreover $\pi_r(X, T_2 \circ T_1) \leq 2\pi_{p_1}(X, T_1)\pi_{p_2}(Y, T_2)$.

If $X = E$ and $Y = F$ then $\pi_r(T_2 \circ T_1) \leq \pi_{p_1}(T_1)\pi_{p_2}(T_2)$.

**Proof**

By Proposition VII.5 it suffices to deal with the case when $1/p_1 + 1/p_2 \leq 1$. We apply Lemma VII.10 with $v = T_1, q = p_1, p = p_2$ and $1/r = 1/p_1 + 1/p_2$; then $1 \leq r < p_1$.

For every finite sequence $x_1, , , x_n$ in $X$ with $M_r((x_i)) \leq 1$, we can write $T_1(x_i) = \sigma_i.y_i$ as in Lemma VII.10; since $r/p_1 + r/p_2 = 1$ it follows from Holder's Inequality that:

$$(\sum_1^n \|T_2 \circ T_1(x_i)\|^r)^{1/r} = (\sum_1^n \sigma_i^r \|T_2(y_i)\|^r)^{1/r} \leq (\sum_1^n \sigma_i^{p_1})^{1/p_1}(\sum_1^n \|T_2(y_i)\|^{p_2})^{1/p_2}$$

Thus, if $m$ is a Pietsch measure for $T_1$, we get

$$\pi_r(X, T_2 \circ T_1) \leq 2m(1)^{1/p_1} \pi_{p_2}(Y, T_2)$$

the result follows by Theorem VII.4.

When $X = E$ and $Y = F$ the proof is the same.

From the Composition Theorem we deduce the following result concerning normal cones:

## VII.12 Theorem

Let $X \subset B$ be a normal cone and $I_B : B \to B$ be the identity map; the following three properties are equivalent:

1) $I_B$ is $p$-summing on $X$ for some $1 \leq p < \infty$.

2) $I_B$ is 1-summing on $X$.

3) There exists $L \in X^0$ and $C > 0$ such that, for every $x \in X$, one has $L(X) \leq \|x\| \leq C.L(x)$.

**Proof**

1) $\Rightarrow$ 2): since $(I_B)^n = I_B$ for every integer $n$, we apply Theorem VII.11 with $n \geq p$; 2) $\Rightarrow$ 1) is obvious.

2) $\Rightarrow$ 3) follows from Theorem VII.7 and the converse is obvious.

## VII.13 Example

The positive cone $X$ of an $L$-space satifies the condition of the preceding theorem.

Another example, due to A. Goullet de Rugy [24] is obtained by taking for $B$ the Banach space of all Radon measures on $[0, 1]$ and for every $0 \leq a \leq 1$ the cone $X_a \subset B$ generated by the set $\epsilon_a - B_1$; for details we refer to our book ([14], II.5.34).

In the preceding examples the sets $X \cap B_1$ and $X_a \cap B_1$ are not relatively compact for the norm topology; we shall see in Theorem VII.16 that, if the identity map on a Banach $B$ is $p$-summing on $B$, then $B_1$ is relatively compact and thus, by a classical theorem, $B$ is finite dimensional.

# VII.14 Proposition

Let $T$ be a $p$-summing operator, $1 \leq p < \infty$, from $B$ to another Banach space; then:

1) $T$ is completely continuous in the sense that, for every sequence $(f_n)$ in $B$ with $f_n \to 0$ for $\sigma(B, B')$, one has $T(f_n) \to 0$ (in norm).

2) $T$ is weakly compact in the sense that $T(B_1)$ is weakly relatively compact.

**Proof**

1) We use the Factorization Theorem VII.6 with the same notations: $(f_n)$ can be viewed as a sequence of continuous functions on the compact set $B'_1$ converging to 0 (at every point); thus $(f_n)$ is uniformly bounded and it suffices to apply the Lebesgue Theorem to conclude that $f_n \to 0$ in $L^p$; the result follows.

2) We can assume that $p > 1$; the proof runs as in 1); it suffices to note that the unit ball of $\mathcal{C}(B'_1)$ is weakly relatively compact in $L^p$.

# VII.15 Proposition

If $T_1 : E \to F$ is a weakly compact operator and if $T_2 : F \to G$ is a completely continuous operator then $T = T_2 \circ T_1$ is a compact operator in the sense that $T(E_1)$ is norm relatively compact.

**Proof**

Let $(f_n)$ be a bounded sequence in $E$; since $T_1$ is weakly compact, by the Eberlein-Smulian Theorem, $T_1(f_n)$ has a subsequence $T_1(f_{n_k})$ converging weakly in $F$; the result follows since $T_2$ is completely continuous.

We are now in a position to state the following result, which follows easily from VII.14 and VII.15..

# VII.16 Theorem

If the identity map on $B$ is $p$-summing on $B$, for $1 \leq p < \infty$, then $B$ is finite dimensional.

Now, we shall study operators whose transposed maps are $p$-summing on a convex cone; these results will be useful in Chapter VIII; when no cone is involved the following lemma is due to S. Kwapien [36] and L. Schwartz [64]. Recall that a set $A$, contained in a vector lattice $L$, is said to be *order bounded* when it is contained in an interval of $L$, namely $A \subset [-x, x]$ for some $x \in L$.

An operator $T$, from a Banach space $B$ to a Banach lattice, is said to be *order bounded* if $T(B_1)$ is order bounded.

# VII.17 Lemma

Let $U$ and $V$ be two Banach spaces and $Z \subset V$ be a closed convex cone. For every $1 \leq p < \infty$ and every $u : U \to V$ the following two properties are equivalent:

1) $u'$ is $p$-summing on $Z^0$ (the polar cone of $Z$).

2) For every $L^p$ space and every positive map $v : V \to L^p$ (in the sense that $v(Z) \subset L^p_+$) $v \circ u$ is order bounded by an element of norm less than $C.\|v\|$, where $C$ is some constant.

Moreover, in 1) one has $\pi_p(Z^0, u') \leq C$ and in 2) one can take $C = 2\pi_p(Z^0, u')$.

**Proof**

1) $\Rightarrow$ 2): suppose $L^p(\Omega, \mathcal{A}, m)$ and $v$ are given. It suffices to prove that, for every finite sequence $x_1,,, x_n$ in $U_1$ and for every measurable partition $A_1,,, A_n$ of $\Omega$, if

$$f = \sum_1^n |1(A_i) v \circ u(x_i)|$$

satisfies $\|f\|_p \leq 2\|v\|\pi_p(Z^\circ, u')$. One has

$$\|f\|_p = \Big(\int \sum_1^n |v \circ u(x_i) 1(A_i)|^p dm\Big)^{1/p} = \Big(\sum_1^n (\int v \circ u(x_i)(\omega) g_i(\omega) m(d\omega))^p\Big)^{1/p}$$

where each $g_i \in L^{p'}$ is of norm 1 and supported by $A_i$.

One has $\|f\|_p = \sum_1^n a_i \langle v \circ u(x_i), g_i \rangle$ where $(a_i)$ is a convenient norm 1 sequence in $\ell^{p'}$. Hence

$$\|f\|_p = \sum_1^n a_i \langle u(x_i), v'(g_i) \rangle = \sum_1^n a_i \langle x_i, u' \circ v'(g_i) \rangle$$

and then $\|f\|_p \leq (\sum_1^n \|u' \circ v'(g_i^+)\|^p)^{1/p} +$ same term with $g_i^-$.

Since the functions $g_i$ are pairwise disjoint, it is easy to see that $(g_i^+)$ is scalarly $\ell^p$ in $L^{p'}$ and that $M_p((g_i^+)) \leq 1$; since $v \geq 0$, one has $v' \geq 0$ and by hypothesis 1) we get $\|f\|_p \leq 2\|v\|\pi_p(Z^\circ, u')$.

2) $\Rightarrow$ 1): let $z_1',,, z_n' \in Z^\circ$ be an arbitrary finite sequence; define $v : V \to \ell^p_n$ by $v(z) = (z_i'(x))$; it is easy to check that $v$ is positive and that $\|v\| = M_p((z_i'))$. By hypothesis 2) there exists $a = (a_1,,, a_n) \in \ell^p_n$ with $\|a\|_p \leq C M_p((z_i))$ such

that $|v \circ u(x)| \leq a$ for every $x \in U_1$; hence $|z_i'(u(x))| \leq a_i$ for $1 \leq i \leq n$, which gives $|\langle u'(z_i), x\rangle| \leq a_i$ and $\|u'(z_i)\| \leq a_i$. It follows that

$$\Big(\sum_1^n \|u'(z_i)\|^p\Big)^{1/p} \leq CM_p((z_i'))$$

which is the result we are aiming for.

In the preceding lemma, if we take $Z = 0$ we have $Z^\circ = V'$ and we retrieve the classical theorem of S. Kwapien [36] and L. Schwartz [64]; in this case the proof shows that we can replace $2\pi_p(u')$ by $\pi_p(u')$.

Using the same proof as in the preceding theorem we shall get the direct part of the following result.

## VII.18 Proposition

Let $1 \leq p < \infty$ and let $u : L^{p'}(\Omega, \mathcal{A}, m) \to B$ be a continuous linear operator. Then:

1) If $u$ is $p$-summing on $L^{p'}_+$, then $u'$ is order bounded.

2) If $u'$ is order bounded, then $u$ is 1-summing on $L^{p'}_+$.

**Proof**

1) $\Rightarrow$ 2): we mimic the proof of 1) $\Rightarrow$ 2) in the above theorem: it suffices to prove that, for every finite sequence $\ell_1,,,\ell_n$ in $B_1'$ and for every measurable partition $A_1,,,A_n$ of $\Omega$ (with obvious notations), the function

$$f = \sum_1^n |1(A_i)u'(\ell_i)|$$

is norm uniformly bounded. We leave the details to the reader.

2) $\Rightarrow$ 1): there exists $f_0 \in L^p$ such that $|u'(\ell)| \leq f_0$ for every $\ell \in B_1'$, hence $|\langle u'(\ell), g\rangle| \leq \langle f_0, g\rangle$ for every $g \in L^{p'}_+$ and $\langle \ell, u(g)\rangle \leq \langle g, f_0\rangle$, thus $\|u(g)\| \leq \langle g, f_0\rangle$ which gives the result.

## VII.19 Remark

The preceding proposition shows that if $u$ is $p$-summing on the positive cone of $L^{p'}$, then it is in fact 1-summing on this cone; this is a result due to Blasco ([15]).

Now, we shall define and study the class of *p-integral* operators, $1 \leq p \leq \infty$; in the preceding Chapters we have met summing operators but not integral operators; the results we shall give in the remainder of this Chapter will be useful in Chapter VIII. They are classical; we mentioned them in order to keep this book self contained; we do not try to give a complete theory but to insist on the results which will be useful in Chapter VIII.

## VII.20 Definition

Let $E$ and $F$ be two Banach spaces. If $1 \leq p \leq \infty$, an operator $T : E \to F$ is said to be *p-integral* if there exist a probability space $(\Omega, \mathcal{A}, P)$, a map $u : E \to L^\infty(P)$, a map $v : L^p(P) \to F''$ such that, if $i_p : L^\infty(P) \to L^p(P)$ is the canonical imbedding, then:

$$T = v \circ i_p \circ u$$

where $F$ is viewed as a subspace of $F''$.

We set $\iota_p(T) = \inf \|u\| \|v\|$, where the infimum is taken over all these factorizations of $T$.

We denote by $\mathcal{I}_p(E, F)$ the set of all these operators.

It is immediate that, if $1 \leq p \leq q \leq \infty$, then $\mathcal{I}_p(E, F) \subset \mathcal{I}_q(E, F)$ and $\iota_q(\ ) \leq \iota_p(\ )$. Moreover $\mathcal{I}_p(E, F) \subset \Pi_p(E, F)$ and $\pi_p(\ ) \leq \iota_p(\ )$. The converse is not true, but it is not immediate to give an example of an operator which is $p$-summing ($p \neq 2$) but not $p$-integral; we refer to ([20], 5.13).

Before going further let us indicate several examples of integral operators.

## VII.21 Examples

1) Every $p$-summing operator from a $\mathcal{C}(K)$ space is $p$-integral and the two norms are equal; it suffices to apply Theorem VII.4.

2) Every $p$-summing operator with values in an $L^\infty(m)$ space, where $m$ is finite, is $p$-integral and the two norms are equal, because $L^\infty(m)$ is an injective space, namely, for every Banach space $V$ and every subspace $W \subset V$ every operator from $W$ to $L^\infty(m)$ extends to all $V$ with the same norm. (It suffices to mimic the proof of the classical Hahn-Banach Theorem given in I.1).

3) Every 2-summing operator is 2-integral and the two norms are equal: with the notations of Theorem VII.6, it suffices to use the projection from $L^2(m)$ onto its closed subspace $F$.

4) For $1 \leq p < \infty$, every order bounded operator (see before VII.17) $T : B \to L^p(\Omega, \mathcal{A}, m)$, where $m$ is arbitrary, is $p$-integral; namely, if $T(B_1) \subset [-f, f]$, with $f \in L^p(m)$, then $\iota_p(T) \leq \|f\|$:

   It suffices to consider the map $u(x) = T(x)/f \in L^\infty(P)$, where $P$ is the probability $(f^p/\|f\|^p)m$, and the map $v : L^p(P) \to L^p(m)$ defined by $v(g) = fg$; one has $\|v\| = \|f\|$ and thus $\iota_p(T) \leq \|f\|$.

5) If $T$ is $p$-integral, with obvious notations the same thing is true for $v \circ T \circ u$ and $\iota_p(v \circ T \circ u) \leq \|u\| \|v\| \iota_p(T)$.

# VII.22 Remark

Contrary to the case of $p$-summing operators we are not able to modify the definition of $p$-integral operators to deal with cones. Consequently, the results we shall give in the remainder of this Chapter are classical and we shall give sometimes sketched proofs.

It is possible to give an equivalent definition of integral operators as in Theorem VII.6,3) concerning summing operators; we admit this result whose proof involves the preceding proposition.

# VII.23 Theorem

Let $E$ and $F$ be two Banach spaces and $1 \leq p \leq \infty$; for every $T : E \to F$ the following two properties are equivalent:

1) $T$ is $p$-integral.

2) For every norming compact set $K \subset E'_1$, there exist a probability Radon measure $P$ on $K$, a map $v : L^p(P) \to F''$ such that, if we denote by $u$ the canonical embedding from $E$ to $\mathcal{C}(K)$ and by $i_p$ the canonical embedding from $L^\infty(P)$ to $L^p(P)$, for every $x \in E$ one has:

$$T(x) = v \circ i_p \circ u(x)$$

where $E$ is viewed as a subspace of $E''$.

Moreover, for every $K$, then $\iota_p(T) = \inf \|v\|$ over all possible $P$ and $v$.

We admit also the following Theorem.

## VII.24 Theorem

For every Banach spaces $E$ and $F$ and $1 \leq p \leq \infty$ then $\iota_p(\ )$ is a norm on $\mathcal{I}_p(E, F)$, which is a Banach space for this norm.

Generally, the transposed map of a $p$-summing operator is not $p$-summing; for $p$-integral operators the same thing is true except if $p = 1$; indeed we have the following theorem which is important for Chapter VIII.

## VII.25 Theorem

For every Banach spaces $E$ and $F$ and every $T : E \to F$ (linear continuous) the following two properties are equivalent:

1) $T$ is 1-integral.

2) $T'$ is 1-integral.

Moreover $\iota_1(T) = \iota_1(T')$.

**Proof**

It is easy noticing that $F' \subset F'''$ and that $L^1(P)$ is hereditary in $(L^\infty(P))'$ (see Lemma VI.10).

Let us explain why the class of 1-integral operatord is important for us; if $(a_{i,j})$ is a square $n \times n$ matrix, by definition the *trace* of $(a_{i,j})$ is $tr((a_{i,j})) = \sum_1^n a_{i,i}$; in fact this definition admits an intrinsic form: if $E$ is an $n$ dimensional vector space, for any (linear) map $A : E \to E$ the number $\sum_1^n \langle A(e_i), e_i^* \rangle$, where $(e_i)$ is a basis of $E$ and $(e_i^*)$ its conjugate basis, does not depend on the chosen basis; if $(a_{i,j})$ is the matrix of $A$ in this basis it is clear that:

$$\sum_1^n \langle A(e_i), e_i^* \rangle = tr((a_{i,j}))$$

Consequently, we can denote this number by $tr(A)$; when things make sense one has: $tr(AB) = tr(BA)$.

The following lemma is due to A. Grothendieck [26], [27], or ([20], 6.14).

## VII.26 Lemma

Let $E$ be a finite dimensional Banach space; for every linear map $A : E \to E$ one has:

$$|tr(A)| \leq \iota_1(A)$$

**Proof** (sketch)

For simplicity (with the notations of VII.20), we deal with the particular case when $\Omega = \{1, 2, , , k\}$ and $P = \sum_1^k p_i.\epsilon_i$: one has $A = v \circ i_1 \circ u$ and $tr(A) = tr(u \circ v \circ i_1)$; if $\varphi_i$ is defined on $\Omega$ by $\varphi_i(j) = \delta_{i,j}$ one has $\|u \circ v \circ i_1(\varphi_i)\| \leq p_i \|u\| \|v\|$, thus $|tr(A)| \leq \|u\| \|v\|$.

The general case involves some technical details; we refer to ([20], 6.14).

The following theorem will be very useful in the case when $1/p + 1/q = 1$; it is due to A. Persson and A. Pietsch [53].

## VII.27 Theorem

Let $1 \leq p, q < \infty$ such that $1/r = 1/p + 1/q \leq 1$ and $E, F, G$ be three Banach spaces. If $u : E \to F$ and $v : F \to G$ are two linear operators then:

1) If $u$ is $p$-summing and $v$ is $q$-integral, then $v \circ u$ is $r$-integral and $\iota_r(v \circ u) \leq \pi_p(u)\iota_q(v)$.

2) if $v$ is $p$-integral and $v$ is $q$-summing, then $v \circ u$ is $r$-integral and $\iota_r(v \circ u) \leq \iota_p(u)\pi_q(v)$.

**Proof** (sketch)

1) Using the definitions, there is a Radon probability $P$ on $E'_1$ and a Radon probability $Q$ on $F'_1$ and a $p$-summing map from $E \subset \mathcal{C}(E'_1)$ to $L^\infty(Q)$; by VII.21,2) this map can be extended to the whole of $\mathcal{C}(E'_1)$ (with the same $\pi_p$ norm). Now it suffices to compose with the canonical map $L^\infty(Q) \to L^q(Q)$ and to apply Theorem VII.11.

2) The proof is similar.

## VII.28 Notes and remarks

1) From an historical point of view, 1-integral operators were introduced by A. Grothendieck in 1955, ([25], [26], [27]) in the framework of tensor products; namely if $E$ and $F$ are two Banach spaces let $E \otimes_\epsilon F$ be the space $E \otimes F$ equipped with the norm of uniform convergence on $E'_1 \times F'_1$ (the *injective topology*). Using Hahn-Banach's Theorem, it is easy to see that for every $\varphi \in (E \otimes_\epsilon F)'$ with $\|\varphi\| = 1$, there exists a Radon probability $P$ on $E'_1 \times F'_1$ (equipped with the duality with $E \times F$) such that for every $x \in E$ and $y \in F$:

$$\varphi(x, y) = \int \langle x, u \rangle \langle y, v \rangle P(du, dv)$$

$\varphi$ can be viewed as a linear map $T : E \to F'$ such that $\varphi(x,y) = \langle T(x), y\rangle$; let us prove that this map is 1-integral:

a) Define $U : E \to L^\infty(P)$ by $U(x)(u,v) = \langle x, u\rangle$ for every $x \in E$ and $(u,v) \in E'_1 \times F'_1$.

b) For every $f \in L^1(P)$ let $proj(fP)$ be the projection of the measure $fP$ on $F'_1$ (namely $proj(fP)$ is the Radon measure on $F'_1$ defined by $H \mapsto \int H(v)\varphi(u,v)P(dudv)$ for every continuous function $H$ on $F'_1$).

c) For every Radon measure $m$ on $F'_1$ let $b(m) \in F'$ be the barycenter of the measure $m$ (for every $y \in F$ one has $\langle y, b(m)\rangle = \int \langle y, v\rangle m(dv)$).

d) Let $V : L^1(P) \to F'$ defined by $f \mapsto b(proj(fP))$.

If $i_1$ is the canonical map $L^\infty(P) \to L^1(P)$ it is easy to see that $T = V \circ i_1 \circ U$; in other words $T$ is 1-integral from $E$ to $F'$.

2) If $T : E \to F$ is $p$-integral and takes its values into a closed subspace $G \subset F$ then $T$ needs not to be $p$-integral as a map into $G$. However, if $T : E \to F$ is $p$-summing, and if $I$ denotes the unit sphere of $F'$ and $i$ the canonical map from $F$ to $\ell^\infty(I)$ then $i \circ T$ is $p$-integral.

In the case when $p = 1$ the class of 1-summing maps were called "applications semi-intégrales à droite" by A. Grothendieck in [25] and "applications préintégrales à droite" in [26], [27].

3) In 1967 A. Pietsch [54] introduced $p$-summing operators, proving Theorems VII.4 and VII.6.

4) In 1971 A. Pietsch [55] introduced $p$-integral operators; previously A. Persson and A. Pietsch [53] had introduced the class of *strictly p-integral* operaors, namely in Definition VII.20 $v$ takes its values in $F$. It is not easy to give an example of a $p$-integral operator which is not strictly $p$-integral (see [20] p. 121).

5) 1-summing operators on normal cones were studied by B. Walsh ([67] page 132) in 1973. Within the framework of Banach lattices, variations in notions related to $p$-summing operators were introduced (probably for the first time) in 1969 by U. Schlotterbeck ([63] Chapters 4.3 and 4.4); there are also the papers of L.P. Yanovskii [69], of N.J. Nielsen and J. Szulga [50] , and of J. Szulga [65]; likewise, O. Blasco published several papers in this domain (for example see [15], [16]). More recent works are those of N. Danet [18] and O.I. Zhukova [70].

6) Theorem VII.4 was proved in [7], [9]; Theorem VII.11 was proved in [11]; Lemma VII.10 is similar to Lemma 2.23 in [20] (implicit in [54]) but with a different proof; indeed, in the proof of Lemma 2.23 in [20], Pietsch's

Factorization Theorem is used but, in our case, we have only at our disposal Pietsch's Inequality of Theorem VII.4.

# Chapter VIII

# Tensor products and positive maps

In this Chapter, $E$ and $F$ will be two Banach spaces and $X \subset E$, $Y \subset F$ will be two closed generating convex cones (not necessarily proper). We first study some norms on the tensor product $E \otimes F$ involving the cones $X$ and $Y$. The results obtained in the preceding Chapters enable us to compare these norms and to study some classes of operators.

Before describing these norms let us indicate what they are not: they depend not only on $E$ and $F$ but also on $X$ and $Y$ and so we have no satisfactory theory concerning morphisms between Banach spaces; in other words our norms are not "functors".

When $X = E$ and $Y = F$, our norms are nothing but the classical Lapreste norms [37], which are generalizations of the Chevet-Saphar tensor norms ([17], [61]).

Then, if $U$ and $V$ are two Banach spaces, and $\Gamma \subset V$ is a closed normal cone, for every $1 < s < \infty$ we determine the structure of all maps $\psi : U \to V$ such that, for every $L^{s'}$ space $B$ and every positive map $T : V \to B$ the map $T \circ \psi$ is integral.

## VIII.1 Definition

If $1 \leq p, q, r \leq \infty$ satisfy $1/r + 1/p' + 1/q' = 1$, we define for every $z \in E \otimes F$:

$$\alpha^+_{p,q}(z) = \inf\{\|(\lambda_i)\|_r M_{q'}((x_i)) M_{p'}((y_i)) : z = \sum_1^n \lambda_i x_i \otimes y_i, x_i \in X,\ y_i \in Y\}$$

# VIII.2 Proposition

1) $\alpha^+_{p,q}$ is a norm on $E \otimes F$.

2) If $p_1 \leq p_2$ and $q_1 \leq q_2$ then $\alpha^+_{p_2,q_2} \leq \alpha^+_{p_1,q_1}$.

3) There exists $m > 0$ such that, for every $x \in E$ and $y \in F$, one has:

$$m\|x\|\|y\| \leq \alpha^+_{p,q}(x \otimes y) \leq \|x\|\|y\|$$

**Proof**

1) It is clear that $\alpha_{p,q}(z) = \alpha_{p,q}(-z)$ for every $z \in E \otimes F$.

For the triangular inequality let $z_1, z_2 \in E \otimes F$; if $\epsilon > 0$ is given, we can choose two decompositions $z_j = \sum_1^n \lambda_i^j x_i^j \otimes y_i^j$, $x_i^j \in X$, $y_i^j \in Y$, where $j = 1, 2$, such that:

$$\|(\lambda_i^j)\|_r \leq (\alpha^+_{p,q}(z_j) + \epsilon)^{1/r}$$

$$M_{q'}((x_i^j)) \leq (\alpha^+_{p,q}(z_j) + \epsilon)^{1/q'}$$

$$M_{p'}((y_i^j)) \leq (\alpha^+_{p,q}(z_j) + \epsilon)^{1/p'}$$

We build a decomposition of $z = z_1 + z_2$ by gluing the sequences $(\lambda_i^1)$ and $(\lambda_i^2)$, $(x_i^1)$ and $(x_i^2)$, $(y_i^1)$ and $(y_i^2)$. Using the form of Holder's Inequality used in the proof of Lemma VII.10 (with $r, q', p'$) the triangular inequality follows.

2) If $1/p_1 + 1/q_1 = 1$ there is nothing to prove, hence we assume $r_1 < \infty$ (see Definition VIII.1); we set:

$$1/p = 1/p_1 - 1/p_2 \text{ and } 1/q = 1/q_1 - 1/q_2$$

consequently: $r_1/r_2 + r_1/p + r_1/q = 1$.

Let $z \in E \otimes F$; if $\epsilon > 0$ is given, we choose a decomposition $z = \sum_1^n \lambda_i x_i \otimes y_i$, $x_i \in X$, $y_i \in Y$ such that:

$$\|(\lambda_i)\|_{r_1} M_{q_1'}((x_i)) M_{p_1'}((y_i)) \leq (1 + \epsilon)\alpha^+_{p_1,q_1}(z)$$

Wet set $\epsilon_i =$ (sign of) $\lambda_i$ and $\mu_i = |\lambda_i|$; let us consider the following decomposition:

$$z = \epsilon_i \mu_i^{r_1/r_2} (\mu_i^{r_1/q} x_i) \otimes (\mu_i^{r_1/p} y_i)$$

One has:

$\|(\lambda_i^{r_1/r_2})\|_{r_2} = \|(\lambda_i)\|_{r_1}^{r_1/r_2}$

$M_{q_2'}((\mu_i^{r_1/q} x_i)) \leq \|(\lambda_i)\|_{r_1}^{r_1/q} M_{q_1'}((x_i))$ by Holder's Inequality with $q/q_2'$ and $q_1'/q_2'$, noticing that $q_2'/q + q_2'/q_1' = 1$.

$M_{p_2'}((\mu_i^{r_1/p} y_i)) \leq \|(\lambda_i)\|_{r_1}^{r_1/p} M_{p_1'}((y_i))$ by Holder's Inequality with $p/p_2'$ and $p_1'/p_2'$, noticing that $p_2'/p + p_2'/p_1' = 1$.

It follows that $\alpha^+_{p_2,q_2}(z) \leq (1 + \epsilon)\alpha^+_{p_1,q_1}(z)$

3) If $x \in E$ and $y \in F$ it is clear that $\|x\|\|y\| \leq \alpha^+_{p,q}(x \otimes y)$ using the definition of $\alpha^+_{p,q}$ and considering $x' \in E'_1$ with $x'(x) = \|x\|$ and $y' \in F'_1$ with $y'(y) = \|y\|$. If $x \in X$ and $y \in Y$ it is again clear that $\|x\|\|y\| \geq \alpha^+_{p,q}(x \otimes y)$; this finishes the proof, since $X$ and $Y$ are closed and generating.

To compare the various norms we have defined it is necessary to determine the dual space of $E \otimes F$ for these norms. We consider first the case $q = 1$.

# VIII.3 Proposition

The norm $\alpha^+_{p,1}$ does not depend on $X$ (up to equivalence).

Moreover, $\alpha^+_{1,1}$ is equivalent to the projective norm $\pi$ defined by

$$\pi(z) = \inf\{\sum \|x_i\|\|y_i\| : z = \sum x_i \otimes y_i\}$$

where the infimum is taken over all the finite representations of $z$ and $x_i \in E$ and $y_i \in F$.

**Proof**

Let $\beta = \beta^+_{p,1}$ be the norm defined as $\alpha = \alpha^+_{p,1}$ where $X$ is replaced by $E$; then $\beta \leq \alpha$; conversely, let $z \in E \otimes F$; for every $\epsilon > 0$ we can write $z = \sum_1^n \lambda_i x_i \otimes y_i$ with $x_i \in E_1$ and $y_i \in Y$ such that: $\alpha(z) \leq \|(\lambda_i)\|_p M_{p'}((y_i)) + \epsilon$.

Since $X$ is closed and generating, by Klee's Theorem (I.15) we have $x_i = u_i - v_i$ with $u_i, v_i \in X$ and $\|u_i\| + \|v_i\| \leq C\|x_i\|$ where $C$ is some constant, thus $z = \sum_1^n \lambda_i u_i \otimes y_i - \sum_1^n \lambda_i v_i \otimes y_i$; the result follows.

The result concerning $\alpha^+_{1,1}$ is proved in the same way.

We are now in a position to determine the dual space of $E \otimes F$ for $\alpha^+_{p,1}$ when $X = E$.

# VIII.4 Proposition (with $X = E$)

If $1 < p \leq \infty$, the dual space of $E \otimes F$ for $\alpha^+_{p,1}$ is identical to the space $\Pi_{p'}(Y, F, E')$ of all the operators $T : F \to E'$ which are $p'$-summing on the cone $Y$, and the dual norm is equivalent to $\pi_{p'}(Y, \ )$.

Namely, if $\phi \in (E \otimes F)'$ there exists a unique $T \in \Pi_{p'}(Y; F, E')$ such that, for all $x \in E$ and $y \in F$: $\phi(x \otimes y) = \langle x, T(y)\rangle$.

**Proof**

By Proposition VIII.2 it is immediate that $\phi$ can be viewed as a continuous map $T : F \to E'$. If $z = \sum_1^n \lambda_i x_i \otimes y_i$ one has $\phi(z) = \sum_1^n \lambda_i \langle x_i, T(y_i)\rangle$.

Since $\alpha^+_{p,1}(z)$ is evaluated by taking $\|(\lambda_i)\|_p = 1$, $x_i \in E$ of norm 1 and $y_i \in Y$ such that $M_{p'}((y_i)) \leq \alpha^+_{p,1}(z) + \epsilon$ the result follows.

To compare the various norms $\alpha^+_{p,1}$ when $X = E$ with the same cone $Y \subset F$, we shall use the index $i(Y)$ defined in IV.1.

## VIII.5 Proposition (with $X = E$)

If $p > i(Y)$ the two norms $\alpha^+_{p,1}$ and $\alpha^+_{\infty,1}$ are equivalent; namely:

$$\alpha^+_{\infty,1}(z) = \inf\{M_1((y_i)) : z = \sum_1^n x_i \otimes y_i,\ x_i \in E,\ \|x_i\| \leq 1,\ y_i \in Y\}$$

**Proof**
It suffices to apply IV.13 and IV.15.

Now we shall deal with the general case when $p \neq 1$ and $q \neq 1$. We denote by $G^*$ the algebraic dual of the vector space $G$. We assume that $X \subset E$ and $Y \subset F$ are two closed convex generating cones.

We admit the following result of calculus.

## VIII.6 Lemma

If $1 < p, q < \infty$ satisfy $1/p + 1/q = 1$, for every $a, b \in \mathbb{R}^+$ one has:

$$ab = \inf\{a^p t^p/p + b^q t^{-q}/q : t > 0\}$$

We are now in a position to prove the following.

## VIII.7 Theorem ($p \neq 1, q \neq 1; 1/r + 1/p' + 1/q' = 1$)

For $\phi \in (E \otimes F)^*$ the following four properties are equivalent:

1) $\phi \in (E \otimes F)'$ for $\alpha^+_{p,q}$.

2) There exist two probability Radon measures $\lambda$ on $E'_1$ and $\mu$ on $F'_1$ and a constant $C$ such that for every $x \in X$ and every $y \in Y$ one has:

$$|\phi(x \otimes y)| \leq C(\int |x(u)|^{q'} \lambda(du))^{1/q'} (\int |y(v)|^{p'} \mu(dv))^{1/p'}$$

3) There exist two Banach spaces $U$ and $V$ and two operators $u : E \to U$ and $v : F \to V$, such that $u$ is $q'$-summing (on $E$) and $v$ is $p'$-summing (on $F$), and, for every $x \in X$ and every $y \in Y$ one has:

$$|\phi(x \otimes y)| \leq \|u(x)\| \|v(y)\|$$

4) Same assertion as in 3) where we only require $u$ to be $q'$-summing on $X$ and $v$ to be $p'$-summing on $Y$.

Moreover, in 2) the best possible constant $C$ is the norm of $\phi$; in 3) the infimum of $\pi_{q'}(u)\pi_{p'}(v)$ is the norm of $\phi$; in 4) the infimum of $\pi_{q'}(X, u)\pi_{p'}(Y, v)$ is the norm of $\phi$.

**Proof**

1) $\Rightarrow$ 2): for evey finite sequences $\lambda_1,,,\lambda_n \in \mathbb{R}$, $x_1,,,x_n \in X$, $y_1,,,y_n \in Y$ one has:

$$|\sum_1^n \lambda_i \phi(x_i \otimes y_i)| \leq \|\phi\| \|(\lambda)\|_r M_{q'}((x_i)) M_{p'}((y_i))$$

It follows that:

$$(\sum_1^n |\phi(x_i \otimes y_i)|^{r'})^{1/r'} \leq \|\phi\| M_{q'}((x_i)) M_{p'}((y_i))$$

Since $r'/p' + r'/q' = 1$, by Lemma VIII.6 the above inequality is equivalent to:

$$\sum_1^n |\phi(x_i \otimes y_i)|^{r'} \leq \|\phi\|^{r'} ((r'/q') M_{q'}^{q'}((x_i)) t^{q'/r'} + (r'/p') M_{p'}^{p'}((y_i)) t^{-p'/r'})$$

for every $t > 0$.

In fact, replacing $x_i$ by $t^{-1/r'} x_i$ and $y_i$ by $t^{1/r'} y_i$ we can assume that $t = 1$ in the above formula; we are now in a position to apply Lemma II.4; in fact we imitate the proof of Theorem VII.4, using the above formula (with $t = 1$) and dealing with the compact product space $K = E'_1 \times F'_1$.

Indeed, with the preceding notations, we consider the family of functions on $K$ defined by:

$$(u, v) \mapsto \|\phi\|^{r'} ((r'/q') \sum_1^n |x_i(u)|^{q'} + (r'/p') \sum_1^n |y_i(v)|^{p'}) - \sum_1^n |\phi(x_i, y_i)|^{r'}$$

This family of functions (where finite sequences $x_1,,,x_n \in X$ and $y_1,,,y_n \in Y$ vary) is a convex cone (notice that, if $x_i$ is replaced by $t^{1/q'}x_i$ and $y_i$ by $t^{1/p'}y_i$, then $\phi(x \otimes y)$ is replaced by $t^{1/r'}\phi(x \otimes y)$) and each of them takes on a positive value; consequently, there exists a probability Radon measure $P$ on $K$ such that, for every $x \in X$ and $y \in Y$ one has

$$|\phi(x \otimes y)|^{r'} \leq \|\phi\|^{r'}((r'/q') \int |x_i(u)|^{q'} P(dudv) + (r'/p') \int |y_i(v)|^{p'} P(dudv))$$

Let $\lambda$ be the projection of $P$ on $E_1'$ and $\mu$ be the projection of $P$ on $F_1'$ ($\lambda(A) = P(A \times F_1')$ and $\mu(A) = P(E_1' \times A)$) ; the result follows by Lemma VIII.6 replacing $x_i$ by $t^{1/r'}x_i$ and $y_i$ by $t^{-1/r'}y_i$.

2) $\Rightarrow$ 3): it suffices to take $U = L^{q'}(\lambda)$ and for $u/C$ the canonical embedding; we take $V = L^{p'}(\mu)$ and for $v$ the canonical embedding.

3) $\Rightarrow$ 4): this is immediate.

4) $\Rightarrow$ 1): if $z = \sum_1^n \lambda_i x_i \otimes y_i$, with $x_i \in X$ and $y_i \in Y$ then $|\phi(z)| \leq \sum_1^n |\lambda_i| \|u(x_i)\| \|v(y_i)\|$; since $1/r + 1/p' + 1/q' = 1$, by the form of Holder's Inequality used in the proof of Lemma VII.10 we have:

$$|\phi(z)| \leq \|(\lambda_i)\|_r (\sum_1^n \|u(x_i)\|^{q'})^{1/q'} (\sum_1^n \|v(y_i)\|^{p'})^{1/p'}$$

Since $u$ is $q'$-summing on $X$ and $v$ is $p'$-summing on $Y$ the result follows.

The assertions concerning the various constants hold by the way things were set up.

# VIII.8 Proposition

1) If $q' > i(Y)$, then for all $p_1, p_2$ such that $i(Y) < p_1, p_2 \leq q'$, the two norms $\alpha^+_{p_1,q}$ and $\alpha^+_{p_2,q}$ are equivalent on $E \otimes F$.

2) If $1/i(X) + 1/i(Y) > 1$, for every $q_1, q_2 > i(X)$ and $p_1, p_2 > i(Y)$ the two norms $\alpha^+_{p_1,q_1}$ and $\alpha^+_{p_2,q_2}$ are equivalent (when they make sense).

**Proof**

It suffices to apply the preceding theorem and IV.13 and IV.15.

To obtain more precize results concerning the family of norms that we have defined, we assume now that $Y = F$; we recall that in this case $i(F) = q_F$.

## VIII.9 Theorem ($p \neq 1, q \neq 1; 1/r+1/p'+1/q' = 1$)

If $X \subset E$ is a closed generating convex cone and $Y = F$, then for $\phi \in (E \otimes F)^*$ the following three properties are equivalent:

1) $\phi \in (E \otimes F)'$ for $\alpha^+_{p,q}$.

2) There exist a reflexive Banach space $Z$ and two operators $u : E \to Z$ which is $q'$-summing on $X$ and $v : F \to Z'$, which is $p'$-summing such that, for every $x \in E$ and $y \in F$:

$$\phi(x \otimes y) = \langle u(x), v(y) \rangle$$

3) Same condition as 2), but where $Z$ is not necessarily reflexive.

Moreover, $\|\phi\|$ is the infimum of $\pi_{p'}(v)\pi_{q'}(X, u)$ over all the representations in 2) or 3).

**Proof**

1) $\Rightarrow$ 2): we take the notations of Theorem VIII.7; we take for $Z'$ the closure in $L^{p'}(\mu)$ of (the image of) $F$ and for $v$ the canonical inclusion. We can regard $\phi$ as a map $\psi : F \to E'$, namely $\phi(x \otimes y) = \langle x, \psi(y) \rangle$ for every $x \in E$ and $y \in F$. Since $Y = F$, for every $x \in X$ and $y \in F$ we have:

$$|\langle x, \psi(y) \rangle| \leq C(\int |x(s)|^{q'} \lambda(ds))^{1/q'} \|v(y)\|$$

Consequently, $\psi$ can be written $\psi(y) = \widetilde{\psi}(v(y))$, where $\widetilde{\psi}$ is a map from $Z'$ to $E^*$ since $X$ is generating; for every $x \in X$ and $z \in Z'$ we have:

$$|\langle x, \widetilde{\psi}(z) \rangle| \leq C(\int |x(s)|^{q'} \lambda(ds))^{1/q'} \|z\|$$

Since $X$ is generating it follows from this inequality and Klee's Theorem (I.15) that $\widetilde{\psi}$ is continuous and takes its values in $E'$; it suffices to take for $u$ the restriction of $(\widetilde{\psi})'$ to $E$.

2) $\Rightarrow$ 3) is obvious.

3) $\Rightarrow$ 1): it suffices to mimic the proof of 4) $\Rightarrow$ 1) in Theorem VIII.7.

The assertion concerning $\|\phi\|$ follows from the way things were set up.

In the framework of the preceding theorem, it is clear that every $\phi \in (E \otimes F)'$ can be viewed as a map $\psi : F \to E'$ as it is indicated in the proof of this theorem; we shall try to characterize these maps $\psi$; for this purpose we shall assume that $r = \infty$ and that $E$ is reflexive. We shall need the results concerning *trace* and integral operators mentioned in Chapter VII.

# VIII.10 Theorem ($p \neq 1, q \neq 1, 1/p' + 1/q' = 1$)

If $E$ is reflexive and $X \subset E$ is a closed generating convex cone, and if $Y = F$, then for every $\psi : F \to E'$ the following three properties are equivalent:

1) $\psi \in (E \otimes F)'$ for $\alpha^+_{p,q}$. Moreover, one has $\|\psi\| \leq C$, where $C$ is defined in 3).

2) There exist a reflexive Banach space $A$ and two operators $v : F \to A$ which is $p'$-summing and $w : A \to E'$, with $w' : E \to A'$ $q'$-summing on $X$, such that $\psi = w \circ v$. Moreover, one has: $\|\psi\| = \inf\{v_{p'}(v)\pi_{q'}(X, w')\}$ over all such representations of $\psi$.

3) There exists $C > 0$ such that for every $L^{q'}$ space $B$ and for every continuous linear map $T : E' \to B$ which is positive on $E'$ (ordered by $X^0$), $T \circ \psi$ is integral and $\iota_1(T \circ \psi) \leq C\|T\|$. Moreover one can take $C = 2\|\psi\|$.

**Proof**

1) $\Rightarrow$ 2): with the notations of (the beginning of) the proof of Theorem VIII.9 we take $A = Z'$ and $w = \widetilde{\psi}$; we have $\psi = w \circ v$; since $Z$ and $E'$ are reflexive the result follows.

2) $\Rightarrow$ 3): from Lemma VII.17 it follows that $T \circ w$ is order bounded by an element of norm less than $2\pi_{q'}(X, w')\|T\|$ and so, by VII.21.4), this map is $q'$-integral, from $A$ to $B$ and $\iota_{q'}(T \circ w) \leq 2\pi_{q'}(X, w')\|T\|$.

Since $v$ is $p'$-summing, it follows from Theorem VII.27 that $T \circ \psi = (T \circ w) \circ v$ is 1-integral and that $\iota_1(T \circ \psi) \leq 2\pi_{p'}(v)\pi_{q'}(X, w')\|T\|$.

3) $\Rightarrow$ 1): let $x_1,,,x_n$ be a finite sequence in $X$ and $y_1,,,y_n$ be a finite sequence in $F$; we define a map $T_1 : E' \to \ell^{q'}_n$ by $x' \mapsto (x'(x_i))$ and a map $T_2 : \ell^{q'}_n \to F$ by $(a_i) \mapsto \sum_1^n a_i y_i$; we have $\|T_1\| = M_{q'}((x_i))$ and $\|T_2\| = M_{p'}((y_i))$.

The map $\theta = T_1 \circ \psi \circ T_2$ satisfies $\iota_1(\theta) \leq \iota_1(T_1 \circ \psi)\|T_2\|$ and by condition 2) one has $\iota_1(\theta) \leq C\|T_1\|\|T_2\|$. It is easy to check that $trace(\theta) = \sum_1^n \langle x_i, \psi(y_i)\rangle$; the relation $trace(\theta) \leq \iota_1(\theta)$, proved in VII.26, yields:

$$\sum_1^n \langle x_i, \psi(y_i)\rangle \leq C M_{q'}((x_i)) M_{p'}((y_i))$$

Consequently, $\psi$ is continuous on $E \otimes F$ for $\alpha^+_{p,q}$ and the result follows.

# VIII.11 Remark

To apply IV.13 and IV.15, we need to have $1/q_F + 1/i(X) > 1$. Since $q_F \geq 2$ this implies that $i(X) < 2$; of course this is impossible if $X = E$ because in that case $i(E) = q_E \geq 2$.

The following theorem, where no tensor product is involved, can be deduced from Theorem VIII.10 when the space $V$ is assumed to be reflexive. However, we shall avoid this hypothesis, giving a direct proof.

## VIII.12 Theorem

Let $U$ and $V$ be two Banach spaces, and $\Gamma \subset V$ be a closed convex normal cone. If $1 < s < \infty$, for a continuous linear map $\psi : U \to V$ the following three properties are equivalent:

1) There exist a reflexive Banach space $W$, two maps $\alpha : U \to W$ and $\beta : W \to V$ such that $\psi = \beta \circ \alpha$, where $\alpha$ is $s$-summing and $\beta'$ is $s'$-summing on the cone $\Gamma^0 \subset V'$.

2) Same condition as 1) but $W$ is not assumed to be reflexive.

3) There exists $C > 0$ such that for every $L^{s'}$ space $B$ and every continuous linear map $T : V \to B$, which is positive on $\Gamma$, $T \circ \psi$ is integral and $\iota_1(T \circ \psi) \leq C\|T\|$.

Moreover, in 1) and 2) one has $\inf\{\pi_s(\alpha)\pi_{s'}(\Gamma^0, \beta')\} \leq C$, where the inf is taken over all the involved representations and in 3) one can take $C = 2\inf\{\pi_s(\alpha)\pi_{s'}(\Gamma^0, \beta')\}$.

**Proof**

1) $\Rightarrow$ 2) is obvious; let us prove that 2) $\Rightarrow$ 3): the arguments are the same as in the proof of 2) $\Rightarrow$ 3) of Theorem VIII.10. $T \circ \beta$ is order bounded in $B$ by an element of norm less than $2\pi_{s'}(\Gamma^0, \beta')\|T\|$, and so it is $s'$-integral with $\iota_{s'}(T \circ \beta) \leq 2\pi_{s'}(\Gamma^0, \beta')\|T\|$ ; consequently, by Theorem VII.27, $T \circ \psi = T \circ \beta \circ \alpha$ is integral and $\iota_1(T \circ \beta \circ \alpha) \leq 2\pi_s(\alpha)\pi_{s'}(\Gamma^0, \beta')$.

3) $\Rightarrow$ 1): if $u_1, , , u_n \in U$ and $v_1', , , v_n' \in \Gamma^0 \subset V'$, we define two maps $\theta_1 : \ell_n^{s'} \to U$ by $(a_i) \mapsto \sum_1^n a_i x_i$ and $\theta_2 : V \to \ell_n^{s'}$ by $v \mapsto (v_1'(v), , , v_n'(v))$ which is positive on $\Gamma$. As in 3) $\Rightarrow$ 1) of Theorem VIII.10, we see that $\theta_2 \circ \psi \circ \theta_1$ is 1-integral and satisfies:

$$\iota_1(\theta_2 \circ \psi \circ \theta_1) \leq CM_s((u_i))M_{s'}((v_i'))$$

it follows from Lemma VII.10 that:

$$\sum_1^n \langle \psi(u_i), v_i' \rangle \leq CM_s((u_i))M_{s'}((v_i'))$$

As in the proof of Theorem VIII.7, we see that there exist two probability Radon measures $\lambda$ on $U_1'$ and $\mu$ on $V_1$" such that for every $u \in U$ and $v' \in \Gamma^0$:

$$|\langle \psi(u), v' \rangle| \leq C(\int |u(x)|^s \lambda(dx))^{1/s}(\int |v'(y)|^{s'} \mu(dy))^{1/s'}$$

Let $W$ be the closure of (the image) of $U$ in $L^s(\lambda)$ and $\alpha$ be the canonical map from $U$ to $W$; one has $\pi_s(\alpha) \leq 1$. If $\alpha(u) = 0$ then $\langle \psi(u), v' \rangle = 0$ for every $v' \in \Gamma^0$ and thus $\psi(u) = 0$ since $\Gamma$ is normal and $\Gamma^0$ is generating by Krein's Theorem (I.12). Consequently there exists a map $\beta$ defined on $\alpha(U)$ such that

$$|\langle \beta(\alpha(u)), v' \rangle| \leq C\|\alpha(u)\|(\int |v'(y)|^{s'} \mu(dy))^{1/s'}$$

We see that $\beta$ is continuous and extends to $W$; by abuse of notations we still denote by $\beta$ this extension; it follows from the above formula that $\beta'$ is $s'$-summing on $\Gamma^0$ and that $\pi_{s'}(\Gamma^0, \beta') \leq C$.

## VIII.13 Remarks

1) By IV.13 and IV.15, $s_1, s_2 > i(\Gamma^0)$ and $s_1', s_2' > q_V$, then the two classes of operators between $U$ and $V$ in the preceding theorem are identical.

2) When $V$ is reflexive, the preceding theorem can be deduced from Theorem VIII.10 letting $F = U,\ E = V',\ X = \Gamma^0 \subset E,\ p = s',\ q = s$.

The following theorem is deduced from Theorem VIII.12 by duality.

## VIII.14 Theorem

Let $G$ and $H$ be two reflexive Banach spaces, and $\Lambda \subset G$ be a closed generating convex cone. If $1 < \sigma < \infty$, for every continuous linear map $\phi : G \to H$ the following two properties are equivalent:

1) There exist a reflexive Banach space $W$, two linear maps $a : G \to W$ and $b : W \to H$, such that $\phi = b \circ a$, where $a$ is $\sigma$-summing on $\Lambda$ and $b'$ is $\sigma'$-summing.

2) There exists a constant $C$ such that, for every $L^{\sigma'}$ space $B$ and every positive continuous linear map $S : B \to G$, $\phi \circ S$ is 1-integral and $\iota_1(\phi \circ S) \leq C\|S\|$.

Moreover, in 1) one has $\inf\{\pi_{\sigma'}(b')\pi_\sigma(\Lambda, a)\} \leq C$ and in 2) one can take $C = 2\inf\{\pi_{\sigma'}(b')\pi_\sigma(\Lambda, a)\}$, where the inf is taken over all possible representations of $\phi$.

**Proof:**

1) $\Rightarrow$ 2): from $\psi = b \circ a$ we deduce $\psi' = a' \circ b'$. If we set $U = H'$, $V = G'$, $\Gamma = \Lambda^0$, $\alpha = b'$, $\beta = a'$ and $s = \sigma'$ we may apply Theorem VIII.12 to $\varphi'$ and $S'$; it follows that $S' \circ \phi'$ is integral and, by Theorem VII.25 the same holds for $\phi \circ S$ with the required estimation.

2) $\Rightarrow$ 1): we use again Theorem VII.25 ; using transposed map it follows from Theorem VIII.12 (with its notations) that $\phi'$ can be written $\phi' = \beta \circ \alpha$ where $\alpha$ is $\sigma'$-summing and $\beta'$ is $\sigma$-summing on the cone $\Lambda \subset G$.

# VIII.15 Notes and remarks

1) When no cone is involved in Definition VIII.1, namely $X = E$ and $Y = F$, the norms we defined are due to J.T. Lapreste [37] and are generalizations of the Chevet-Saphar norms ([17], [61]). The point is that these norms behave well with respect to morphisms between Banach spaces; they are "functors".

   When $p = 1$, $q = 1$ the norm $\alpha_{1,1}$ is the classical projective norm of A. Grothendieck:

   If $z \in E \otimes F$ we have $\alpha_{1,1}(z) = \inf\{\sum_1^n \|x_i\|\|y_i\| : z = \sum_1^n x_i \otimes y_i\}$

   The space $E \otimes F$ equipped with this norm is denoted by $E \otimes_\pi F$. It is called the projective tensor product because, if $E_0$ and $F_0$ are two other Banach space, and $S : E_0 \to E$ and $T : F_0 \to F$ are norm one and onto, then $S \otimes T : E_0 \otimes F_0 \to E \otimes F$ is norm one and onto.

   Another classical norm is the injective norm, denoted by $\| \ \|_\epsilon$:

   If $z \in E \otimes F$ with $z = \sum_1^n x_i \otimes y_i$ then: $\|z\|_\epsilon = \sup\{\sum_1^n \langle x', x_i\rangle\langle y', y_i\rangle : x' \in E_1', \ y' \in F_1'\}$

   The space $E \otimes F$ equipped with this norm is denoted by $E \otimes_\epsilon F$; the norm $\| \ \|_\epsilon$ is not a Lapreste norm. It is called the injective norm because, if $E$ is a subspace of $E_0$ and $F$ is a subspace of $F_0$, and if $S : E \to E_0$ and $T : F \to F_0$ are the canonical injections then $S \otimes T : E \otimes F \to E_0 \otimes F_0$ is an isometry.

   These notions were introduced by A. Grothendieck; we refer to the books by A. Defant and K. Floret [19] and to the one by R.A. Ryan [60].

   A close notion is the one of *ideal of operators*; we refer again to the same books.

2) The norm $\alpha_{2,2}$ is denoted by $w_2$; it is quite important; A. Grothendieck proved the following Theorem:

   For every integer $n$, on $\ell_n^\infty \otimes \ell_n^\infty$ the two norms $\| \ \|_\pi$ and $\| \ \|_{w_2}$ satisfies:

$$\| \ \|_{w_2} \leq \| \ \|_\pi \leq K_G \| \ \|_{w_2}$$

   $K_G$ is called the Grothedieck Constant.

In fact A. Grothendieck proved the equivalent form of this Theorem:

Every continuous linear operator $T : \ell^1 \to \ell^2$ is 2-summing and satisfies:

$$\pi_2(T) \leq K_G \|T\|$$

An equivalent form of these results is the so-called Grothendieck inequality:

For every integer $n$ and every real numbers $a_{i,j}$, where $1 \leq i, j \leq n$, such that

$$\sup\{|\sum a_{i,j} t_i t_j| : |t_i| \leq 1,\ |t_j| \leq 1\} \leq 1$$

for every norm one vectors $x_1, , , x_n$ and $y_1, , , y_n$ in a Hilbert space one has:

$$|\sum a_{i,j} \langle x_i, y_j \rangle| \leq K_G$$

# Appendix A
# Type and cotype

In this Appendix we collect some classical results concerning *type* and *cotype* theory, which is due to B. Maurey and G. Pisier [46]. After the publication of B. Maurey's book [44], B. Maurey and G. Pisier studied series in Banach spaces in the framework of probability theory.

Let $B$ be an infinite dimensional Banach space and $(x_n)$ be a sequence in $B$; if $(\epsilon_n)$ denotes the sequence of Rademacher's functions the problem was to find conditions on $1 \leq r \leq \infty$ such that:

1) If the series $\sum \epsilon_n x_n$ converges (a.e.) in $B$, then the series $\sum \|x_n\|^r$ converges. This is the cotype problem.

2) If the series $\sum \|x_n\|^r$ converges, then the series $\sum \epsilon_n x_n$ converges (a.e.) in $B$. This is the type problem.

To begin with let us recall Dvoretzky's Theorem ([20], 19).

## A.1 Theorem

Let $B$ be an infinite dimensional Banach space. For every integer $n$ and every $\epsilon > 0$ there exist $x_1,,,x_n$ in $B$ such that for every scalars $a_1,,,a_n$:

$$(1-\epsilon)\|(a_i)\|_2 \leq \|\sum_1^n a_i x_i\| \leq (1+\epsilon)\|(a_i)\|_2$$

In their Article, B. Maurey and G. Pisier introduced the following two definitions:

## A.2 Definition

$B$ is said to be of *cotype* $q$ if, for every finite sequence $x_1,,,x_n$ in $B$, one has:

$$(\sum_1^n \|x_i\|^q)^{1/q} \leq C \int \|\sum_1^n \epsilon_i(\omega)x_i\|d\omega$$

where $C$ is some constant.

The *cotype index* $q_B$ of $B$ is the infimum of the $q$ such that $B$ is of cotype $q$.

It is easy to see that $q_B \geq 2$ (use A.1).

## A.3 Definition

Let $B$ be an infinite dimensional Banach space. $B$ is said to be of *type* $p$ if, for every finite sequence $x_1,,,x_n$ in $B$, one has:

$$\int \|\sum_1^n \epsilon_i(\omega)x_i\|d\omega \leq C(\sum_1^n \|x_i\|^p)^{1/p}$$

where $C$ is some constant.

The *typpe index* of $B$ is the supremum of the $p$ such that $B$ is of cotype $p$. It is easy to see that $p_B \leq 2$ (use A.1).

## A.4 Definition

If $1 \leq r \leq \infty$ we say that $B$ contains uniform copies of finite dimensional $\ell^r$ spaces if for every integer $n$ there exist $x_1,,,x_n$ in $B$ such that for every scalars $a_1,,,a_n$:

$$m\|(a_i)\|_r \leq \|\sum_1^n a_ix_i\| \leq M\|(a_i)\|_r$$

where $0 < m \leq M < \infty$ are two constants.

In fact, one can prove that one can require that $m = 1 - \epsilon$ and $M = 1 + \epsilon$.

The following two theorems are due to B. Maurey, G. Pisier and J.L. Krivine [46].

## A.5 Theorem

1) $q_B$ is the supremum of the sets of all $r$ such that $B$ contains uniform copies of finite dimensional $\ell^r$ spaces.

2) $q_B$ is the infimum of the set of all $q$ such that the identity is $(q, 1)$-summing on $B$ in the sense that, for every finite sequence $x_1, , , x_n$ in $B$, one has:

$$(\sum_1^n \|x_i\|^q)^{1/q} \leq CM_1((x_i))$$

where $C$ is some constant.

## A.6 Theorem

$p_B$ is the infimum of the sets of all $r$ such that $B$ contains uniform copies of finite dimensional $\ell^r$ spaces.

The following result is easy:

## A.7 Proposition

If $B$ is of type $p$, then $B'$ is of cotype $p'$.

The following theorem is a deep result due to G. Pisier ([20], 13.17).

## A.8 Theorem

Assume that $B$ satisfies $p_B > 1$; if $B$ is of cotype $q$, then $B'$ is of type $q'$.

## A.9 Notes and remarks

The connection between Maurey's Theory and the above results is Theorem IV.15, with $X = B$ and Theorem A.5.2); it is Proposition I.4 of [46].

Moreover, the proofs of Theorems A.5 and A.6 involve some results of Nikisihn's paper [51], which is one of the roots of Maurey's Theory (see II.10) and the contribution of J.L. Krivine to these theorems is a variation of his proof of Theorem V.11.

For historical comments concerning the above results, with refer to B. Maurey's Article in the Handbook ([45] p. 1299-1332).

# Appendix B
# Concave and convex operators

In the Introduction (§B) (see also V.10) we have recalled what a Banach lattice is; in II.10.5) we have indicated what a "Krivine element" is and we have met $p$-concave and $p$-convex operators; in this Appendix we gather some results (without proofs) concerning the more general families of $(q,p)$-concave operators and of $(p,q)$-convex operators, where $1 \leq p \leq q$, introduced by B. Maurey [43].

## B.1 Definition

Let $L$ be a Banach lattice and $E$ be a Banach space.

A linear operator $u : L \to E$ is said to be $(q,p)$-*concave*, where $1 \leq p \leq q$, when, for every finite sequence $x_1,,,x_n$ in $L$ one has:

$$(\sum_1^n \|u(x_i)\|^q)^{1/q} \leq C\|(\sum_1^n |x_i|^p)^{1/p}\|$$

where $(\sum_1^n |x_i|^p)^{1/p}$ is the "Krivine element".

A linear operator $v : E \to L$ is said to be $(p,q)$-*convex*, where $1 \leq p \leq q$, when, for every finite sequence $x_1,,,x_n$ in $E$:

$$\|(\sum_1^n |v(x_i)|^q)^{1/q}\| \leq C(\sum_1^n \|x_i\|^p)^{1/p}$$

where $(\sum_1^n |v(x_i)|^p)^{1/p}$ is the "Krivine element".

The following theorem is due to B. Maurey [43].

## B.2 Theorem

If $1 \leq p < q$ then $u$ is $(q,p)$-concave iff $u$ is $(q,1)$-concave.

## B.3 Proposition

If $1 < p \leq q$ then:
1) $u$ is $(q,p)$-concave iff $u'$ is $(q',p')$-convex.
2) $v$ is $(p,q)$-convex iff $v'$ is $(p',q')$-convave.

# Appendix C Another proof of Proposition V.8.

In Proposition V.8 we have noticed that, if $X$ and $Y$ are two normal cones (in a Banach space $B$) contained in a same normal cone, then:

$$i(X+Y) = \sup\{i(X), i(Y)\}$$

In this Appendix we prove that $I_X \cap I_Y \subset I_{X+Y}$ which implies obviously the above result using Theorem IV.15. Our proof will rely only on Theorem IV.10,c); we shall need some results concerning conical measures related to the theory of Integral Representation. We shall state these results in the general framework of Hausdorff locally convex spaces (H.l.c.s.).

For convenience we recall the definition of Choquet's Order mentioned in III.24,1.

## C.1 Definition

Let $E$ be a H.l.c.s; we denote by $\prec$ the order on the set of positive conical measures on $E$ defined by:

$$(\lambda \prec \mu) \text{ iff } \lambda(f) \leq \mu(f) \text{ for every } f : x \mapsto \sup\{\ell_1(x),,,\ell_n(x)\},$$

$x \in E$, where $\ell_1,,,\ell_n \in E'$

We state (without proofs) the following easy results.

## C.2 Proposition

1) If $\lambda \prec \mu$ then $r(\lambda) = r(\mu)$.

2) For every $\mu$, if $x = r(\mu)$, one has $\epsilon_x \prec \mu$.

3) For every $x, y \in E$ if $z = x + y$ one has $\epsilon_z \prec \epsilon_x + \epsilon_y$.

The following result is an easy consequence of Lemma III.20.

## C.3 Proposition

Let $X \subset E$ and $Y \subset E$ be two cones in $\mathcal{S}$, contained in a same cone in $\mathcal{S}$; then the sum $Z = X + Y$ is a cone in $\mathcal{S}$.

The following lemma is the key point for our endeavour.

## C.4 Lemma (with the notations of A.3)

For every $\mu \in M^+(Z)$, there exists $\mu_X \in M^+(X)$ and $\mu_Y \in M^+(Y)$ such that $\mu \prec \mu_X + \mu_Y$.

**Proof**

By Lemma III.10 we may write $\mu = \lim(\sum_i \epsilon_{z_i})$ along some ultrafilter $\mathcal{U}$, where the $\sum_i$ is finite (namely $i$ belongs to some finite set), $z_i \in Z$ and $\sum_i z_i = r(\mu)$; we can write $z_i = x_i + y_i$ where $x_i \in X$ and $y_i \in Y$. By Proposition A.2 one has:

$$\sum_i \epsilon_{z_i} \prec \sum_i \epsilon_{x_i} + \sum_i \epsilon_{y_i}$$

One has $\sum_i x_i \leq r(\mu)$ and $\sum_i y_i \leq r(\mu)$ for the order on $E$ whose positive cone is $Z$.

It follows from Lemma III.20 that $\sum_i \epsilon_{x_i}$ and $\sum_i \epsilon_{y_i}$ converge respectively along $\mathcal{U}$ to conical measures $\mu_X \in M^+(X)$ and $\mu_Y \in M^+(Y)$, for the duality with $h(E)$, and $\mu \prec \mu_X + \mu_Y$.

We are now in a position to prove the result we are aiming for.

## C.5 Proposition

Let B be a Banach space; if $X \subset B$ and $Y \subset B$ are two normal cones, contained in a same normal cone, then $I_X \cap I_Y \subset I_{X+Y}$.

**Proof**

We equip $B$ with an increasing norm for the normal cone $Z = X + Y$; the result follows from Proposition V.2 and the above lemma, noticing that the function $f$ involved in Theorem IV.10,c) is sublinear and consequently, with obvious notations $\mu(f) \leq \mu_X(f) + \mu_Y(f)$.

# Appendix D
# Some problems

In this Appendix we have gathered some problems encountered all along our book.

## D.1 Problem

In IV.3 we have dealt with Sobolev spaces in a trivial case. If $B$ is a Sobolev space and if $X$ is its positive cone (for the order of functions) is there a formula giving $i(X)$? The same question arises for the polar cone $X^0 \subset B'$, which is a normal cone by Klee's Theorem (I.15), since $X$ is closed and generating .

## D.2 Problem

Krivine's Theorem (V.11) provides a formula giving the supremum of all $1 \leq r \leq \infty$ having a nice property. Is there a formula giving the infimum of these $r$ ? May be this formula could be, if $V$ is a Banach lattice:

$$\sup\{p : \|\sum_1^n x_i\| \leq C(\sum_1^n \|x_i\|^p)^{1/p}\}$$

for every finite sequence $x_1,,,x_n$ in $V^+$, with $x_i \wedge x_j = 0$ for $i \neq j$, where $C$ is some constant.

## D.3 Problem

In the case of a normal cone $X \subset B$ a problem, which is similar to the preceding one, is indicated in V.23.

## D.4 Problem

In Chapter IV, when the cone $X \subset B$ is not assumed to be normal, we have no theorem similar to Theorem V.17; the question is made precise in V.22.

# Appendix E
# Historical comments

All my life I have kept alive in myself a strong interest in Mathematics. In this Appendix, I intend to show to the reader that to a large extent, this book results from a combination of circumstances:

After my secondary education at Lycée Buffon in Paris, I moved on to Lycée Louis-le-Grand in Paris for highly competitive further education and our teacher, Mr Berrard, urged us to take the first-year exam at Jussieu University. As I went down from Lycée Louis-le-Grand to Jussieu, I walked past a bookshop which displayed a copy of Riesz-Nagy's book in the window. I bought it immediately.

Some years later, at university, I attended G. Choquet's course, with a title which rang a bell with me: "Topology and the theory of functions". He supervised my doctoral dissertation, and I learned the theory of Integral Representation and Conical Measures.

At the same time, a friend of mine advised me to join the S.M.F. (French Mathematical Society). At that time the Society used to give its members copies of *Astérisque*. I chose the book by B. Maurey as its title meant something to me, and that is how I became aware of Rosenthal's Lemma. I think that this Lemma has never been used since, because this approach has been superseded by the type and cotype theory.

That is why I regard this book as the fruit of a combination of circumstances.

# References


[1] Y.A.Abramovich and C.D.Aliprantis Positive Operators. *Handbook of the Geometry of Banach Spaces* (2001), 85-122.

[2] F.Albiac and N.J.Kalton *Topics in Banach Space Theory.* Graduate Texts in Math., Springer, 2006.

[3] C.D.Aliprantis and O.Burkinshaw *Positive operators.* Academic Press, New York 1985.

[4] T.Ando On fundamental properties of a Banach space with a cone. *Pacific J. Math. 12* (1962), 1163-1169.

[5] S.Banach Sur les fonctionnelles linéaires. *Studia Math. 1* (1929), 211-216.

[6] S Banach *Théorie des opérations linéaires.* Warszawa, 1932.

[7] R.Becker Structure des cônes normaux dans un espace de Banach ou son dual. *C. R. Acad. Sci. Paris Sér. I Math. 314* (1992), 535-539.

[8] R.Becker Structure des cônes normaux dans un espace de Banach ou son dual, II. *Bulletin Austral. Math. Soc. 47-3* (1993), 427-434.

[9] R.Becker Cones contained in a Banach space and factorization of positive operators defined on its dual with values in a space $L^1$. *Math. Zeitschrift 216* (1994), 257-263.

[**10**] R.BECKER On ordered Banach spaces. *Rev. Roumaine Math. Pures Appl. 44, 5-6* (1999), 737-753.

[**11**] R.BECKER On normal cones and summing operators on Banach spaces. *Rev. Roumaine Math. Pures Appl. 68, 4* (2003), 373-383.

[**12**] R.BECKER On Positive Operators on Some Ordered Banach Spaces. *Positivity 9* (2005), 249-257.

[**13**] R.BECKER Convex cones, tensor products and operators. *Rev. Roumaine Math. Pures Appl.* to appear.

[**14**] R.BECKER *Convex cones in analysis.* Hermann, Paris, 2006.

[**15**] O.BLASCO Positive $p$-summing operators on $L_p$ spaces. *Proc. Amer. Math. Soc. 100* (1987), 275-280.

[**16**] O.BLASCO Vector values harmonic functions and cone absolutely summing operators. *Illinois J. Math. 33* (1983), 292-300.

[**17**] S.CHEVET Sur certains produits tensoriels topologiques d'espaces de Banach. *Warscheinlichkeitstheorie verw. Gebeite 11* (1969), 120-138.

[**18**] N.DANET Opérateurs sommants sur le cône positif d'un espace linéaire. *Stud. Cerc. Mat. 46, 5* (1994), 515-520.

[**19**] A.DEFANT AND K.FLORET *Tensor Norms and Operators Ideals.* North Holland Mathematics Studies 176, 1993, Amsterdam.

[**20**] J.DIESTEL, H.JARCHOW AND A.TONGE *Absolutely Summing Opearators.* Cambridge University Press, 1995.

[**21**] E.DUBINSKY, A.PELCZYNSKI AND H.P.ROSENTHAL On Banach spaces $X$ for which $\Pi_2(\mathcal{L}_\infty, X) = B(\mathcal{L}_\infty, \mathcal{X})$. *Studia Math. 44* (1972), 617-648.

[**22**] N.DUNFORD AND J.T.SCHWARTZ *Linear operators, Part I.* Interscience publishers, New York, 1957.

[**23**] S.R.FOGUEL *The Ergodic Theory of Markov Processes.* Van Nostrand Mathematical Studies, Vol. 21, New York, London, Toronto, Melbourne.

**[24]** A.GOULLET DE RUGY Sur les cones engendrés par une famille de convexes compacts. *Bull. Soc. Math. de France 97* (1973), 241-252.

**[25]** A.GROTHENDIECK *Produits tensoriels topologiques et espaces nucléaires.* Mém. Amer. Math. Soc.,16, 1955.

**[26]** A.GROTHENDIECK Résumé de la théorie métriques des produits tensoriels topologiques. *Bol. Soc. Mat. Sao Paulo, 8* (1956), 1-79.

**[27]** A.GROTHENDIECK Résumé de la théorie métriques des produits tensoriels topologiques, *Resenhas, 2(4)* (1996), 401-480.

**[28]** H.HAHN Uber lineare Gleichungen in linearen Raumen. *Journal fur Math. 157* (1927), 214-229.

**[29]** E.HELLY Uber lineare Funktionaloperationen. *Sitzsber, Acad. Wiss. Wien 121* (1912), 265-297.

**[30]** G.JAMESON *Ordered Linear Spaces.* Lectures Notes in Math., 141, Springer, Heidelberg, 1970.

**[31]** V.L.KLEE Boundedness and continuity of linear functionals. *Duke Math. J., 12* (1955), 263-270.

**[32]** I.KLUVANEK AND G.KNOWLES *Vector Measures and Control Systems.* North-Holland Mathematics Notes Studies, Notas de Mathematicas 58, Amsterdam, Oxford, New-York, 1976.

**[33]** M.G.KREIN Propriétés fondamentales des ensembles coniques normaux dans l'espace de Banach. *C.R. (Doklady) Acad. Sci. URSS (N.S.), 28* (1940), 13-17.

**[34]** J.L.KRIVINE Théorèmes de factorisation dans les espaces réticulés. *Ecole Polyt. Palaiseau, Sém. Maurey - Schwartz* 1973/1974, Exp. XXII-XXIII.

**[35]** J.L.KRIVINE Sous-espaces de dimension finie des espaces de Banach réticulés. *Ann. of Math. 104* (1976), 1-29.

**[36]** S.KWAPIEN On a theorem of L. Schwartz and its applications to absolutely summing operators. *Studia Math. 38* (1970), 193-201.

[37] J.T.LAPRESTE Opérateurs sommants et factorisations travers les espaces $L^p$. *Studia Math. 57* (1976), 47-83.

[38] L.LE CAM *Notes on asymptotic methods in statistical decision theory.* Centre de Recherches Mathématiques de Montréal (Canada), 1974.

[39] L.LE CAM *Asymptotics methods in statistical decision theory.* Springer Series in Statistics, New-York, Berlin, London, 1986.

[40] H.LEMBERG Nouvelle démonstration d'un théorème de J.L. Krivine sur la finie représentation de $\ell_p$ dans un espace de Banach. *Israel J. Math. 39* (1981), 341-348.

[41] D.LI AND H.QUEFFELEC *Introduction l'étude des espaces de Banach.* Collection SMF, Paris, 2004.

[42] B.MAUREY Une nouvelle caractérisation des applications (p,q) sommantes. *Ecole Polyt. Palaiseau, Sém. Maurey - Schwartz* 1973/1974, Exp. XII.

[43] B.MAUREY Type et cotype dans les espaces munis de structures locales inconditionelles. *Ecole Polyt. Palaiseau, Sém. Maurey - Schwartz* 1973/1974, Exp. XXIV-XXV.

[44] B.MAUREY *Théorèmes de factorisation pour les opérateurs valeurs dans les espaces $L^p$.* Soc. Math. de France, Astérisque 11, Paris 1974.

[45] B.MAUREY Type, Cotype and K-Convexity. *Handbook of the Geometry of Banach Saces* (2001), 1299-1332.

[46] B.MAUREY AND G.PISIER Séries de variables aléatoires vectorielles indépendantes et propriétés géométriques des espaces de Banach. *Studia Math. 58* (1976), 45-90.

[47] S.MAZUR Uber konvexe Mengen in linearen normierten Raumen. *Studia Math. 4* (1933), 70-84.

[48] P.MEYER-NIEBERG *Banach Lattices.* Springer, Berlin, 1991.

[49] I.NAMIOKA *Partially ordered linear topological spaces.* Mem. Amer. Math. Soc. 24, 1957.

[50] N.J.NIELSEN AND J.SZULGA $p$-lattice summing operators. *Math. Nachr. 119* (1984), 219-230.

[51] E.M.NIKISHIN Resonance theorems and superlinear operators. *Russian math. Surv. 25* (1970), 125-187.

[52] A.L.PERESSINI *Ordered Topological Vector Spaces.* Harper and Row, 1967.

[53] A.PERSSON AND A.PIETSCH $p$-nukleare and $p$-integrale Abbildungen in Banachraumen. *Studia Math. 33* (1969), 19-62.

[54] A.PIETSCH Absolut $p$-summierende Abbildungen in Normierten Raumen. *Studia Math. 28* (1967), 333-353.

[55] A.PIETSCH Adjungierte normierte Operaratorenideale. *Math. Nachr., 48* (1971), 189-211.

[56] F.RIESZ Untersuchungen uber Systeme integrierbarer Funktionen. *Math. Annalen, 69* (1910), 449-497.

[57] F.RIESZ AND B.SZ.NAGY *Lecons d'analyse fonctionnelle.* Gauthier-Villars, Paris, 1954.

[58] M.RIESZ Sur le problème des moments. Troisième Note, *Arkiv fur Math., Astronomi och Fysik, 17* (1923), No 16.

[59] H.P.ROSENTHAL On subspaces of $L^p$. *Ann. of Math. 97* (1973), 344-373.

[60] R.A.RYAN *Introduction to Tensor Products of Banach Spaces.* Springer, London, 2002.

[61] P.SAPHAR Produits tensoriels d'espaces de Banach et classes d'applications linéaires. *Studia math. 36* (1970), 70-100.

[62] H.H.SCHAEFER *Topological Vector spaces.* Springer, Berlin, 1971.

[63] H.H.SCHAEFER *Banach Lattices and Positive Operators.* Springer, Berlin, 1974.

[64] L.SCHWARTZ Applications $p$-radonifiantes et théorèmes de dualité. *Studia math. 38* (1970), 203-213.

[**65**] J.SZULGA On $p$-absolutely summing operators acting on Banach spaces. *Studia Math. 81* (1985), 53-63.

[**66**] E.THOMAS *Integral Representations in Convex Cones.* Publications de l'Université de Groningen No ZW-7703, 1977.

[**67**] B.WALSH Ordered vector sequences spaces and related classes of linear operators. *Math. Ann. 206* (1973), 89-138.

[**68**] P.WOJTASZCZYK *Banach Spaces for Analysts.* Cambridge University Press,1991.

[**69**] L.P.YANOVSKII On summing and lattices summing operators and characterizations of AL-spaces. *Sibirskii Math. Zh. 20* (1979), 401-408.

[**70**] O.I.ZHUKOVA On modified classes of $p$-nuclear, $p$-summable and $p$-integrable operators. *Siberian Math. J. 39, 5* (1998), 894-907.

# Definition index

# Notation index

Imprimé en France par
LIGHTNING SOURCE
www.lightningsource.fr

www.ingramcontent.com/pod-product-compliance
Lightning Source LLC
LaVergne TN
LVHW010051061026
849532LV00004BC/1227
* 9 7 8 2 7 0 5 6 6 7 2 1 4 *